Labs of Our Own

Labs of Our Own

Feminist Tinkerings with Science

Sig/Sara Giordano

RUTGERS UNIVERSITY PRESS
NEW BRUNSWICK, CAMDEN, AND NEWARK, NEW JERSEY
LONDON AND OXFORD

Rutgers University Press is a department of Rutgers, The State University of New Jersey, one of the leading public research universities in the nation. By publishing worldwide, it furthers the University's mission of dedication to excellence in teaching, scholarship, research, and clinical care.

Library of Congress Cataloging-in-Publication Data

Names: Giordano, Sig-Sara, author.
Title: Labs of our own : feminist tinkerings with science / Sig/Sara Giordano.
Description: New Brunswick : Rutgers University Press, [2025] | Includes bibliographical references and index.
Identifiers: LCCN 2024026973 | ISBN 9781978840379 (hardcover) | ISBN 9781978840362 (paperback) | ISBN 9781978840386 (epub) | ISBN 9781978840393 (pdf)
Subjects: LCSH: Science—Social aspects. | Science—Political aspects. | Social justice. | Feminism.
Classification: LCC Q175.5 .G56 2025 | DDC 306.4/5—dc23/eng20240927
LC record available at https://lccn.loc.gov/2024026973

A British Cataloging-in-Publication record for this book is available from the British Library.

Copyright © 2025 by Sig/Sara Giordano

All rights reserved. Laboratory glass icon ornament courtesy of vecteezy.com.

No part of this book may be reproduced or utilized in any form or by any means, electronic or mechanical, or by any information storage and retrieval system, without written permission from the publisher. Please contact Rutgers University Press, 106 Somerset Street, New Brunswick, NJ 08901. The only exception to this prohibition is "fair use" as defined by U.S. copyright law.

References to internet websites (URLs) were accurate at the time of writing. Neither the author nor Rutgers University Press is responsible for URLs that may have expired or changed since the manuscript was prepared.

♾ The paper used in this publication meets the requirements of the American National Standard for Information Sciences—Permanence of Paper for Printed Library Materials, ANSI Z39.48-1992.

rutgersuniversitypress.org

Contents

Prelude: You're Either with Us or against Us: Affective Dissonance and 9/11   vii

Introduction   1

Interlude 1: Serendipity   23

1   (De)constructing DIY Community Biology Labs   27

Interlude 2: If We Knew What We Were Doing   52

2   The Tinkerer as a New Scientific Subject   54

Interlude 3: Learning the Limits of Ethical Debate   70

3   Becoming the Informed Public   73

Interlude 4: Nerd Masculinity   96

4   Feminist Labs of Our Own in Academia?   98

Interlude 5: When the Right Comes to the Defense of Science   119

5   Toward Queer Sciences of Failure   121

Interlude 6: Queer Revolt   134

6   Tinkering as a Feminist Praxis   136

Epilogue   170

Acknowledgments   175
Notes   177
References   187
Index   203

PRELUDE

You're Either with Us or against Us

AFFECTIVE DISSONANCE AND 9/11

A few years ago, I attended a community-building event at my university for faculty who work across academic/nonacademic borders. The conveners asked us to each tell a story about a time when we felt that we did not belong. This book is about how community and belonging is built and mobilized in different claims for laboratory space of one's own. I highlight how defining community borders always creates inclusions and exclusions. Therefore, the story that I shared that night of not belonging as I began my scientific career seems an appropriate place to start.

I began graduate school a couple of weeks before September 11, 2001. One month earlier I officially had changed my permanent residence to somewhere outside New York City for the first time in my life. As I remember it, the news of the attacks came in as I was getting ready to head to school that day. I began monitoring the news, and shortly later I received an email that classes would be canceled for the day.

Across the mainstream political spectrum there was little distinction between Democrat and Republican as the rhetoric of "us" versus "them" was agreed upon. For example, then Senator Hillary Clinton's language in September, "Every nation has to either be with us or against us. Those who harbor terrorists or who finance them, are going to pay a price," was nearly identical to that of then U.S. President George Bush: "Every nation, in every region, now has a decision to make. Either you are with us, or you are with the terrorists."[1]

It became clear that the United States was headed to war, and the room for dissent was significantly narrowed: music stations began censoring songs, critique of the United States by TV personalities was met with firing, not putting up an American flag made your neighbors question you, and not standing for the pledge of allegiance at a sports game came with a new level of vitriol. Arabs and Muslims (and those assumed to be so) were being detained by the government and harassed and murdered by white Americans in retaliation. The rhetoric from government

vii

viii PRELUDE

officials directed to other nations (as highlighted with Senator Clinton and President Bush) was clearly meant as a warning to all of us.

The next time class convened, I felt totally alone in a classroom of over fifty other graduate students taking introductory biochemistry. Before class began, the only acceptable conversation about the political moment seemed to be people posturing to make mention of how close they were to someone who knew someone who died, with the requisite public community mourning and anger. Then the class began, and we got right back to biochemistry. I felt so confused and out of place. Talk of invasion was already on the news, and people were being beaten up in the streets. How could we be sitting here going on with business as usual?

As I struggled in a new city with only my partner to trust, we both searched the internet for other kinds of news and counter discourses. There was no Facebook or Twitter, of course, and websites were still relatively new. Our neuroscience program had an email mailing list that reached hundreds of us at a time—students, faculty, affiliated faculty, and other interested community members. As I was sitting at the computer struggling to find something that made sense, an email supposedly about the Taliban came through on the mailing list from one of our senior faculty members. I quickly read the email and immediately knew it was what we now call "fake news." The email claimed that the Taliban was tattooing numbers on prisoners' arms and constructing concentration camps. This attempt to directly compare the Taliban to the Nazi regime was easy to spot, but this senior scholar did not question the news and did not fact check it before sending it out to hundreds of people.

I decided that this would be the forum for me to share my own political thoughts. I crafted an email critiquing the fake news that had been sent out and focusing on what I saw as a dangerous and extreme nationalism in the United States. I called on us as academics who get paid to think and solve problems to be actively involved in stopping this war, state repression, and extreme nationalism. It would be an understatement to say that this email was not looked upon favorably by my colleagues. I received my first hate mail, and it was pretty vicious. Some senior faculty suggested that I deserved to die for my "treason."

I held on tightly to the few emails from people thanking me. Some faculty caught me in the halls to tell me that I was brave. By the end of the email fiasco, the administrative assistant for the program sent out an apology on my behalf for using the mailing list against its purposes, stating that I was new and did not know better—this was in spite of a passionate meeting in her office in which I told her I did not want her to send such an email, I was not sorry, and the original email forwarding the scam about the Taliban is what should be reprimanded. I also received a Sunday morning call at home from the director of graduate study who implored me to not drop out and assured me that I was welcome despite the hate mail and anger among some faculty. I did decide to stay in the program; however, this began what I describe in the fourth chapter as my eventual "failure" and "defection" from the sciences proper.

Throughout *Labs of Our Own*, I look at how communities are formed through affective attachments. Judith Butler was among those arguing for third ways of

PRELUDE

understanding the "us" versus "them" rhetoric: "I propose to consider a dimension of political life that has to do with our exposure to violence and our complicity in it, with our vulnerability to loss and the task of mourning that follows, and with finding a basis for community in these conditions" (2004, 19). Butler asks us to imagine different ways of finding common cause around violence and mourning in the moments after 9/11, forcing us to examine whose lives are and are not grievable and why from a U.S. context. I did not have Butler's words at the time, but I felt a realignment of community and belonging. The affective dissonance that I experienced of not feeling properly connected to the "us" I was supposed to be mourning and avenging through violence toward "others" created a new kind of community belonging. This other community was one that was passionately dedicated to creating another kind of world: real social justice, an end to gendered racial capitalism, queer liberation, and knowledge practices to match these aims.

This book is grounded in feminist science studies. One of the foundational concepts in feminist science studies is that of situated knowledges (Haraway 1988): that is, understanding knowledge as always produced from somebody and somewhere. Therefore, throughout the text I provide interludes that frame the main content of *Labs of Our Own* within my larger political world.

Labs of Our Own

Introduction

In 2008, I received my PhD in neuroscience. This was the culmination of seven years of postgraduate learning. My science education took place not only in the classrooms and labs of Emory University but in the greater Atlanta area as well: in the streets,[1] in coffee shop meetings, and during late-night kitchen table planning sessions as I became immersed in left activism. My initial post-9/11 antiwar activism led me eventually to feminist, antiracist, anticapitalist, queer, intersex, and disability justice activist communities. I found myself occupying positions across typical university/public borders by being in both academic science communities and leftist activist communities.

As I became more and more involved in an activist life, I thought of my PhD work as my day job and did not think about it as connected to the political work. Through this border-crossing experience, I found out how little each community knew about the other. This lack of knowledge was not equal, however. In the case of academic sciences, we did not know much about what activists said about science because we did not need to know what was being said outside the academy because academic researchers typically have the final say on scientific truths. Activists, on the other hand, understood that scientific truths had an important and often negative impact on how our social world was structured but did not know much about the daily workings of scientists.

I began trying to merge my day job (PhD work) with my activism after reading feminist critiques of science. These initial critiques shed light on how my social justice concerns and science work were already intertwined. For example, colleagues were drawing paychecks from Department of Defense grants, reproducing studies about sex differences, and teaching about recent gay gene research out of context. During this period of re-education, I began to see connections that had always been there. I became more vocal in activist communities about my role in science to see where I could do some *local* activist work by challenging systemic oppression at work. Around this time, some reproductive justice activists approached me to ask if I could help them with basic research that they wanted to conduct on birth control.

It was exciting to think that my training could be useful. However, the activists did not understand that scientists work on really specific topics with specific apparatuses that are not inexpensive or easy to move between. My research was on leg muscles in humans. I did not have the skills or access to the materials or equipment necessary to complete the kind of work they wanted me to do. Although activists may have a sophisticated understanding of the way scientific knowledge is co-produced with cultural beliefs and then cyclically used to defend social stratification based on the idea that our social categories of race, gender/sex, and dis/ability are natural, there is less understanding about how scientists operate day to day.

The problem with not understanding our daily work is that it makes collaboration with activists difficult for sympathetic scientists. Academic scientists do not need to know about activist knowledges because of their epistemic authority; however, activists do not know a lot about how science works because the processes have remained opaque to the public.

I found this lack of understanding about science surprising at first because we regularly hear calls for increased public science literacy from scientists. However, these calls are more often about increasing appreciation and trust for science than attempting to increase the amount of people actively engaged in scientific knowledge-making (Roth and Barton 2004). Science literacy, appreciation, and greater public participation are central to *Labs of Our Own*.

The combination of science education that I received inside and outside of traditional methods led me to a deep desire to be part of democratic science projects. It was impossible to be satisfied going to work in a traditional lab, generating questions with colleagues based on funding opportunities, conducting experiments, and then reporting the results in academic journals. I did not want to simply share our findings with my larger science community; I wanted to do scientific research together with communities across and beyond the borders of the university. I wanted to develop research questions from the top-down instead of bottom-up.

Unfortunately, there are not many opportunities to do community-based, participatory research in neuroscience or the other "hard" sciences. We have seen changes in other fields, such as public health, urban planning, and sociology, that have produced socially relevant, community-based research. However, basic science research has lagged in this regard. I have found through my academic and political journeys that I am not unique nor alone in a desire for more democratic sciences. Democratic science is not institutionalized in the ways other fields that have embraced participatory methodologies may be, and experiments in democratic science are taking place in sites outside of biology proper.

Creating more democratic sciences for those traditionally kept out of science has been my focus since leaving the sciences proper in 2008. The activist communities I mentioned earlier are not the only ones who have pushed for more democratic control of scientific knowledge. In 2010 I was introduced to the field of synthetic biology through a feminist science studies postdoctoral research opportunity. The question that piqued my interest the most was how synthetic biology could be used to open biology to a more collective, participatory practice. Synthetic

INTRODUCTION

biology merges methods from genetics and engineering to manipulate biological organisms more effectively and systematically.[2] Through these manipulations it may be possible to use living systems to produce what has been promised as limitless, renewable products that could be used to solve many human problems. Technological advances that have reduced the cost of processing DNA have opened the field for do-it-yourself (DIY) possibilities.

Would synthetic biology open possibilities for truly participatory science practices that address the exclusions I had become concerned about during my PhD work? Would this be a place for the activists who convinced me there was an important role for scientifically trained feminists in movement work? I did not find immediate links between synthetic biology communities and feminist activist communities. Instead of focusing only on synthetic biology and trying to make activists fit in, I broadened my definition of what counted as democratic science experimentation. By widening the scope of what counts as science beyond formal academic fields, I found activists following traditions set forth by the Black Panther Party and feminist women's health movement in environmental and reproductive justice movement work. I was also able to see that many of us in feminist science studies ourselves have experimented with ways to have our students participate in science outside biology classrooms.

Labs of Our Own analyzes these seemingly disparate sites of democratic science experimentation. Who is pushing for democratic sciences and why? How are all these attempts related? What does it mean for how we structure not only science but our larger society through democratic ideals? Not all democratic sciences are created equally, so how do we navigate the various rhetorics of democratic science, social justice, and scientific revolutions?

Labs of Our Own is about the politics and practices of democratic sciences in the mid-teens of the twenty-first century. By democratic sciences, I mean sciences that are participatory, that are conducted outside and across university borders, that not only claim to be "for us" but are created "by us." The question of who is meant by "us" is crucial across the various sites I interrogate—from physical DIY community biology laboratories to feminist studies departments and classrooms to anticapitalist activists and artists. The implications of how greater participation in science through democratic sciences such as DIY biology is realized (or not) matter not only for those who are calling for it. Attempts to make science more democratic impact—and are impacted by—the structuring of our societies. Whether you have participated in DIY or democratic science spaces or not, you are likely impacted by the debates over science that *Labs of Our Own* traces across these sites.

SCIENCE AND POLITICS

In 2001 I imagined that science and politics were wholly separable; today, Science has taken a prominent role in politics, and your feelings toward science help to determine your political orientation. The debates over COVID-19 are a prime example of this. These alignments did not begin with that novel coronavirus nor with Donald Trump's election as president of the United States in 2016. However, the recent widespread investment in taking political sides on science is notable, and it forms a critical part of the context for varying democratic science projects over the last decade.

In January 2017, the Women's March in Washington, DC, occurred the day immediately after Trump was inaugurated as president. A few months later, in April 2017, many of those same protestors were participating in their own city's March for Science. The March for Science was held throughout the world and largely drew from political bases opposed to the sexism and racism represented by Trump's ascent to power.

We also saw lawn signs such as the one in Figure I.1 begin to show up in cities and liberal suburbs, and many varieties of the signs are available on sites ranging from Etsy to Amazon today.[3] In these signs several political movements are grouped together to represent a coherent political worldview. "Science Is Real" and "Kindness Is Everything" stand out as outliers among the list of slogans based on claiming humanness and/or rights for various social groups who are typically excluded from political rights. Kindness Is Everything seems to orient the political approach as one of acceptance, and Science Is Real seems to include science as a disenfranchised category under attack in some way.

Figure I.1. Lawn sign that reads, "In This House, We Believe: Black Lives Matter, Women's Rights Are Human Rights, No Human Is Illegal, Science Is Real, Love Is Love, Kindness Is Everything." The sign has a black background, and each phrase is in a different color with some variation of fonts between different phrases. Image courtesy of Kristin Joiner and Embolden WI.

INTRODUCTION

The March for Science appeared to have this same analogic meaning. Typically, political marches have been marches to support different groups of people or to target a specific policy such as against war or for health care for all—so what did it mean to march for science? Science does not singularly represent a social category of disenfranchised people, nor a specific policy. This usage of political support for science can perhaps be understood better when paired with another popular yard sign: "Support Science: Defend the Earth" with an image of a globe. With the framing of support for science as necessary for defending the earth, the first sign makes more sense. If we insert "earth" for science, then a March for Earth or a plea to protect the earth makes sense as a political call to care about our environment.

Slogans are, of course, limiting due to their brevity, but they are quite important for framing political debates. Looking at the other slogans more carefully, we have seen critiques from different political contingents for each of them. Many of us in queer circles, for example, are quite critical of the focus on love being universal, natural, and the main issue for queer people (e.g., Bernstein Sycamore 2004; Conrad 2014). The overall framing of kindness as the cure, while well-meaning, might obscure the structural reasons for these deeply historical and economically motivated oppressions. Kindness may also be read as a statement by the property owners placing these yard signs to distinguish them from angry, violent, and/or property-destroying protesters.

Instead of looking at each of these slogans separately, it is important to ask what the juxtaposition of these politics together means.[4] The inclusion of Science on the sign as a stand-in for caring about environmental politics might unintentionally narrow the conversation about the environment. The other movements nodded to on the sign have to do with issues commonly associated with social justice, so we might expect environmental concerns to be conveyed through a message for environmental justice. Environmental justice is largely credited to racial justice activists who have called for attention to environmental racism, to indigenous people who have long called for a sustainable relationship with our natural world but have been excluded from the proper sciences, and to other activists who are engaged in earth defense movements. Science itself, while sometimes used by these activists, has also been criticized for its focus on technofixes (Brault 2017) or for repeating conclusions already reached by activists (Wu 2013). The inclusion of Science also raises questions about the erasure of Science's role in determining the hierarchies of gender and race that make necessary in the first place the slogans that argue for equal rights for women as human and Black lives as important and worthy of protection.

This flattening of what Science is assumes that Science is inherently good and on the side of social justice. It is in this political moment that *Labs of Our Own* asks, How do various democratic science projects shape our understandings of not simply science but our political worlds? How do democratic sciences rely on proximity to social justice ideals and particularly inclusion? And who is imagined to be worthy of inclusion in various democratic sciences?

Methodologies, Political-Intellectual Commitments, and Definitions in *Labs of Our Own*

Labs of Our Own analyzes how continued challenges to scientific, economic, and political power are intertwined. These challenges set off some predictable and unpredictable variations in who are considered legitimate/illegitimate holders of such power. I trace how recent challenges to the authority of traditional sciences result in various forms of democratic sciences. Early chapters highlight the ways democratic DIY science labs reproduce familiar exclusions, demonstrating the flexibility of gendered racial capitalism. At the end of the book, I offer feminist tinkering as a methodology that can interrupt such reproductions. *Labs of Our Own* therefore provides both (1) a careful analysis of all claims for democratic sciences even (or especially) when social justice rhetoric is used and (2) a critical approach to countering gendered racial capitalism's flexibility through a praxis that is perhaps even more flexible! It must be a praxis that holds close the certainty of uncertainty and unknowability together with a deep sense of ethical responsibility to try to create better worlds.

For this first objective, I use ethnographic methods such as participant observation and interviews along with discursive analyses of public-facing materials from websites and media attention to understand how DIY biologists define themselves, what conditions produce such an individual, and what the production of a democratic biologist means in this political moment. I refer to this new scientific subject as the *tinkerer*. Using a feminist methodology that includes self-reflection, I analyze the academic field I call home, feminist science studies. I include feminist science studies in democratic science projects because of (1) the movement between academic and activist projects and (2) the strong focus on the lack of diversity in proper sciences for our field. Therefore, I argue our classrooms and research aim to make sciences more democratic.

For the second objective, tinkering with concepts across disciplines and even across the academy/public border allows me to propose a feminist definition of tinkering. The flexibility that Chela Sandoval (2000) argues we can learn from "the oppressed" opens up decolonial approaches that aim for third ways of knowing.

I use the term *gendered racial capitalism* not to mark a specific kind of capitalism but to call attention to the inextricability of gender, race, and capitalism in our historic moment. Central to capitalism is the accumulation of capital "by producing and moving through relations of severe inequality among human groups" (Melamed 2015). To maintain the unequal relationships required to accumulate capital (e.g., workers/CEOs, renters/landlords) there must be a rationality to why certain groups end up as losers/winners. The historic creation of race (Robinson 1983) and gender (Federici 2004) as we know them as organizing systems of society make these inequalities make sense. Although the idea of natural racial distinctions provide/d rationale for different types of people capable of being owners versus laborers, M Adams argues that gender is also a necessary logic for capitalism to succeed because "without reproductive violence and the controlling of women's bodies, the forcing of heteronormativity, the imposition of the nuclear family to produce workers, social reproduction theory and a host of other things, there couldn't have been the same level of wealth accumulation" (Long 2020). Therefore, where I use capitalism unmarked, I am using it as a shorthand that should still be understood as gendered racial capitalism.

Concepts of race and gender also importantly create group solidarities that obscure the central role of the theft of labor and land in maintaining such a system. For example, racism allows poor white male workers to feel more attachment to white capitalists than to other workers across race and gender. Similarly, gendered divisions of labor require wealthy white women to perform unpaid labor and stay beholden to men in heteropatriarchal family structures rather than work across class and race for full control of one's labor. Attending to race and gender together allows for understanding both the specific ways each concept organizes society and the ways these social categories of difference intersect.

Science has been central to stabilizing ideas of supposedly natural divisions of race and gender. Science, as we generally use the term, refers to a culturally specific methodology and set of knowledge claims that are made through a belief in universal, pure objectivity. This universal science became globalized through Western colonization by replacing, taking from, and devaluing other forms of knowledge (Quijano 2000). At times I emphasize this point by using capital-S Science to signal the taken-for-granted claim that Western science is the best way for producing knowledge about ourselves and our worlds. This leaves open the possibility for lowercase and plural sciences to refer to other ethnosciences and for us to understand that the kind of science put forth by Western science is just one ethnoscience in itself (Harding 2001). Similarly, I use capital letter forms of other institutions to call attention to their dominant status and their cultural specificity.

At the same time, this science provided the justification for domination. The development of Western science was crucial in the shift away from theological primacy in Europe to the idea of "humans" as rightful controllers of their own destiny through their ability to reason (Wynter 2003). Through colonization, Science developed as the most rational form of knowing, splitting reason and emotion from practices of knowing and thereby grounding Western science in ideas of objectivity and absolute truths (Anzaldúa [1987] 2012, 99). While Europeans and propertied men became known as rational, on the other side of the binary the Colonized, Women, and Others were labeled as irrational/emotional. This labeling provided justification for why the latter were not allowed to participate in political decision-making, thereby often losing control over their own bodies at an individual level and land on both individual and collective levels.[5] The idea of different types of people, with difference locatable in bodies and passed on to offspring through reproduction, became solidified through Biology in the late nineteenth century. Therefore, sciences and particularly biologies are critical in understanding how inequalities have been maintained and the possibilities for self-determination.

The power of these explanations for inequalities cannot hold if they remain stagnant. This is where the flexibility of gendered racial capitalism comes in. A major confrontation with the taken-for-granted racial and gendered ordering took place globally in the 1960s/70s. Challenges were mounted, showing the deeply unfair ways that race and gender—and relatedly nation—determined control over resources and life itself. With the curtains pulled back, capitalism itself was threatened.

Biology and control over science and medicine were a major part of how gendered racial capitalism was challenged by activists. From the Black Panther Party's health activism to the feminist women's health movement, challenging definitions of nature(s) and health and sickness was integral to their fights for self-determination. These movements called for control over knowledge production itself along with access to the materials needed to survive (Murphy 2012; Nelson 2011). Scientific racism and the role of biological determinism in both gender and race came under sustained fire. The Black Panther Party provided immediate health care for its base by opening community clinics with each chapter; they also called attention to the fallacy of racist biological determinism—for example, the connections made between Blackness and violence (Nelson 2011). Feminist health clinics opened up across the United States to change protocols of care while feminist manuals to rewrite knowledge about women's bodies were distributed, such as the Boston Women's Health Book Collective's *Our Bodies, Ourselves* (Murphy 2012).

The intersection of these political, economic, and biological challenges to what it meant to be human and what that meant for control over oneself and society were met with adjustments, concessions, and co-optation to maintain gendered racial capitalism's dominance (Wynter 2003). This is not the first time capitalism has shown its flexibility in response to its power being threatened.[6] The most notable shift under this threat was the inclusion of certain members of disenfranchised groups into the proper demos—in literal terms with the passing of the Voting Rights Act (1965) and more broadly in positions of political and economic power.

Anticolonial movements became co-opted with Western nations helping to prop up dictators who might still espouse anticolonial rhetoric but in action worked to expand the reach of gendered racial capitalism while they individually benefited. Continued global capitalist domination demanded exceptions even within socialist nation-states to create a global set of citizenship distinctions between workers and the owning classes, which are arguably more important than national citizenship status (Ong 2006). In response to demands for control over knowledge production, Universities (the formal institutional seat of Knowledge) developed programs in women's studies, ethnic studies, Black studies, and other area studies (Wallerstein 2004), increased enrollment of students of color (Fabricant and Brier 2016), began to integrate critiques of eurocentrism into certain fields/curricula (McDonald 1996), and in some cases created more community access and control over research (Wachelder 2003).

These inclusions led to a shift in—but not eradication of—capitalist losers/winners. At the same time, the former, more broad-sweeping, social safety net programs began to disintegrate. For example, some U.S. public universities that had been free or low-cost opened their doors to a more diverse crowd at the same time as they began to increase tuition, thereby creating new disadvantages (Fabricant and Brier 2016). Rationales of innate, biological differences that had led to racial inequalities were replaced by social critiques of the Black family; these critiques specifically placed blame (and control) on Black women and their reproduction while offering inclusion for those who conformed to white, heteronormative versions of the Family (Hong 2015a, b; Roberts 1997). This form of inclusion into capitalism with a hyperfocus on the individual along with removal of public infrastructure characterizes neoliberalism. There are many definitions, but I follow the insights from ethnic studies and queer studies that focus on how this social dimension of inclusion has been key to obscuring the continued differential life chances produced by capitalism (Duggan 2003; Hong 2015b).

Each of the movements *Labs of Our Own* examines developed anticapitalist demands for control over biological truths. How did the challenges put forth by activists involved in the Black Panther Party and feminist women's health movement influence our current demands for more democratic sciences? How did the larger societal context—rebellion and co-optations together—set the stage for how we think about science, truth, and democracy today?

One key change has been a move from the rationality side of the colonially produced rational/emotional binary to an embrace of an emotional approach to Science. This movement is key for understanding calls for greater participation in the sciences and how various forms of more democratic sciences play out. In the case of proving scientific potential, ideas of innate intelligence have been largely sidestepped with a greater focus on being passionate about Science. Similarly, the goal of science literacy has been changed from one focused on educating the "ignorant" public, now derided as a deficit view approach, to that of increasing Science appreciation in the general public (Lewenstein 2003).

A central theme in *Labs of Our Own* is how different understandings of affect open and foreclose possibilities for knowing. Affect is key to neoliberal flexibility. By moving from one side of the rational/emotional binary to the other, affect operates in producing belonging in and exclusion from proper scientific communities. As in the general case of neoliberalism shifts, there may be some changes to who is included, while at the same time a more individual focus leading to entrepreneurial sciences emerges. Thereby, the links between capitalism, science, and coloniality are not broken but strengthened. A decolonial understanding of this binary construction involves recognizing that moving from one side to the other will not solve the fracture created by the initial splitting of reason/emotion. Instead, we will need to understand reason and emotion as intermeshed concepts that are not found outside a given context. Affect and feeling, similar to how we understand truth and knowing, are co-produced in our social, political worlds (Ahmed 2010).

Tinkerers, Tinkerings, and Racialized Underdog Politics

On September 1, 2011, *Science* published a news story, "A Lab of Their Own," to document the creation of the first physical DIY synthetic biology spaces (Kean 2011).

The article's title plays on Virginia Woolf's famous 1929 essay *A Room of One's Own* in which she addressed the impact of gender inequality on women writers. The subject in the *Science* article, however, is not "women" but rather "tinkerers": in this apparent co-optation, there is an extrapolation of social justice and feminist rhetoric to the imagined unmarked subject of the tinkerer. Feminists inside and outside the academy have also laid claim to labs of our own. In particular, feminists trained as scientists, including me, have expressed longing for a return to the lab on our own terms in our own spaces. In 2005, feminist science studies scholar Banu Subramaniam played on Woolf's title as well when she laid out an argument for doing feminist science in "Laboratories of Our Own: New Productions of Gender and Science." More recently, in January 2017, this play on words was used by the *Chronicle of Higher Education* for its article "A Lab of Her Own," which focused on how different colleges attempt to retain women in engineering and computer science majors (Anft 2017).

In each case, the use of Woolf's title frames the problem as one of inequity in access to science. My own work, *Labs of Our Own*, examines how this claim is mobilized across disparate groups: tinkerers and feminists inside and outside academia. I propose to reclaim the label of tinkerer, already prominent in synthetic biology communities (Calvert 2013), for the possibilities it contains for feminist theorizing.

To tinker can mean to mess with or play with something. This can signal an innocence or lack of sophistication. And although tinkering takes place with an intent to make something better, we know that sometimes people tinker with things they ought to have left alone. During the tinkering or proposed tinkering, there are those who may caution against it—"just don't tinker with that anymore" juxtaposed with the tinkerer who is adventurous and determined that he can make it work. This is a gendered script: arguably, tinkering is a masculine act, and further is one of national pride for the United States, relegating those who caution against tinkering as nagging, uninitiated, and antipatriotic—as standing in the way of Progress. This binary is not simply invoked with tinkering as good: some should tinker, and some should not. Some have "no business tinkering with" some things. Who gets to tinker and the bounds of appropriate tinkering are worth analyzing from a feminist perspective. In focusing on *who* tinkers, I am not asking *who* in a demographic sense but rather in the sense of what kind of subject positions, figures, and identities are produced. These formations ultimately pool together in familiar patterns of racial and gendered inclusion/exclusion.

The *Oxford English Dictionary*'s definition of the verb *tinker* begins by stating that it is "in all senses usually depreciative," and the definition of *tinkerer* follows as "one who tinkers or works at mending something in a clumsy or ineffective way."[7] Considering these negative connotations, why would DIY biologists or anyone use tinkering to describe their activities? The combination of co-opted social justice discourse with a U.S. protestant work ethic and love for underdog politics creates a specific American brand of innovation that makes the tinkerer identity make sense. In the United States this reappropriation and love for tinkering crosses

INTRODUCTION   11

liberal and more conservative national politics. A popular history of tinkering by Alec Foege (2013), *The Tinkerers: The Amateur, DIYers, and Inventors Who Make America Great*, locates tinkering in the DNA of America. As the description of the book on its flap reads, "From its earliest years, the United States was a nation of tinkerers: men and women who looked at the world around them and were able to create something genuinely new from what they saw. Guided by their innate curiosity, a desire to know how things work, and a belief that anything can be improved, amateurs and professionals from Benjamin Franklin to Thomas Edison came up with the inventions that laid the foundations for America's economic dominance." The introduction to the book describes current day tinkerers as possessing innate curiosity, desire/passion, and creativity, and they are positioned as fighting against the restraints of big corporate powers.

This resurgence of a populism that is rooted in an anti-intellectual U.S. protestant work ethic can be seen ironically as both a backlash to and co-optation of the racial and gender justice movements. The fear of co-optation of movements is always present—and for good reason. As I referred to earlier, Sylvia Wynter (2003) describes "The Sixties" as a brief challenge to what she calls the "Overrepresentation of Man" as human. She and other critical scholars have noted that the 1960s represented a moment of intersection between important international anticapitalist movements that pointed out the sexist, racist, and homophobic structures that created and maintained dominance over most of the world's inhabitants. These discourses showed the ways that only some humans (predominately white property-owning males) counted as human. These movements largely became co-opted and contained as rights-based claims for "inclusion." In this way they became severely limiting in the scheme of social justice. One co-optation of those movements can be seen as the emergence of neoliberalism (Hong 2015b). Perhaps ironically, another co-optation could be identified in the explicitly racist backlash against neoliberalism that has also been framed in terms of the need to protect a disenfranchised, racialized population: white men.

Neoliberalism (with its various and often vague definitions) has co-opted multiculturalism, feminism, and antiracism at the same time as maintaining a racialized (yet changing) system of economic domination (Duggan 2003). In fact, this incorporation/co-optation has defined a current postfeminist, postracial Western society (particularly in the United States) where these values are assumed to be self-evident and definitional of Western democracy, thereby discounting the need for any further antiracist, feminist activism and also using this assumption to distinguish the West from the rest (Gill 2008).

At the same time, what appears to be the converse of this politics, recently labeled "Trumpism," also relies on identity politics and acts to maintain capitalism. This movement appropriates identity politics through its reverse racism claim and maintains that capitalism itself is not the problem but that instead the distribution of winners and losers is not following the correct racialized lines. The title of Foege's (2013) book suggests it is tinkerers "who make America great," and the resurgence of the tinkerer should be understood in the context of these nationalist politics. During the 2016 U.S. presidential election, multibillionaire Donald Trump rallied an overwhelmingly large white base under the slogan "Make America Great Again." The idea of a better time in American history is read by many as at least in part referring to the good old days of slavery and/or a pre-civil rights era and includes nostalgia for old-fashioned family values that include "proper" roles for women and children and the centrality of a heterosexual unit. Tied to the cultural/racial politics of this statement is the idea that there was more economic prosperity (for white men) during this period.

In the current moment, part of what this movement addresses is the impact of globalization on white American laborers. Although globalization has had detrimental effects on much of the world's population, framing the problem as one of "foreigners" taking jobs (in the United States and abroad) splits the working class and poor people by producing a racialized enemy for the white male. Because liberals and conservatives across the United States' two political parties expected to be celebrating narrowly avoiding Making America Great Again, the result of the election of Donald Trump in 2016 seemingly shocked many capitalists. Capitalists who are concerned that Trump's extremism is dangerous for capitalism's continued dominance strategize about how to deal with this political reorganization, but one thing is probably safe to assume: keeping the discourse of a legitimate racial stratification viable (but in check) is important to maintaining economic inequality through racial separation (racial capitalism). The movement for what I call "new" democratic sciences in the teens of the twenty-first century emerges from this political moment.

Tinkerer labs have popped up in science museums around the world as a fun, playful, and accessible way to increase public interest in science. Although tinkering has had widespread appeal as a way to make science accessible to all, the tinkerer seems to be a racialized subject. Similar acts of what would look like tinkering can be interpreted either as diligent scientific or technological curiosity or as dangerous, terrorist activity, depending on the actor. The tinkerer in dominant narratives plays on a populist idea of the everyday man having a kind of ingenuity by working with his hands. This figure is imagined most as a white working-class man (even though this likely does not match the actual middle/upper-class status of many tinkerers who are taking up various new hobbies working with their hands). The way the tinkerer as an acceptable figure is positioned against the terrorist or criminal can be seen in cases such as that of Ahmed Mohamed, who was fourteen years old when he made international news after bringing a homemade clock to school and was detained by school authorities and then the police under fears of "terrorism." His experimental invention was not deemed to be acceptable tinkering arguably because of his race and religion, highlighting the intersections between Islamophobia, U.S. imperialism, and the school-to-prison pipeline (Kazi 2015). In response, some wishing to support him reframed his work in terms of "tinkering" (e.g., Higgins, Wesmacott, and Uzielli 2015). However, Nazia Kazi (2015) suggests this correction in the form of one student's inclusion as a good kind of tinkerer misses the structural racialized roots that led to the initial incident.

INTRODUCTION 13

In Foege's analysis of American tinkering, economic success or entrepreneurialism is the natural consequence of healthy tinkering. *Entrepreneur*, similar to the category of tinkerer, describes racially acceptable forms of economic independence in the United States today. There are drastic differences in the narratives and consequences surrounding entrepreneurial enterprises such as Uber that work around the law for profit versus Black men such as Eric Garner and Alton Sterling who sold loose cigarettes (Baker, Goodman, and Mueller 2015) and CDs (Lane 2016), respectively, on the street. In the case of the two men, the consequence was death at the hands of police; in the case of Uber, public forums have held slow debates of regulations only after massive individual and industry protests and alongside at least an equal admiration for the ingenuity and copying of such entrepreneurship (Chan and Kwok 2021).

Labs of Our Own brings this political understanding to bear on the figure of the *biological* tinkerer. How do biological tinkerers position themselves within debates about scientific curiosity and bioterrorism? What might be the impacts of carving out space for legitimate biological tinkering? How does the tinkerer make use of underdog politics?

Bioeconomies, Affective Circulations, and Interventions

Biological tinkerers are typically positioned in opposition to Big Bio and powerful, bureaucratic government or university projects. Their labs are typically found in old warehouses where new startups (legal and illegal) are renting out space. A small group of committed volunteers who are often former biotech researchers and current graduate students form the dedicated core that makes the physical space a possibility. The feeling in these spaces is casual and informal. It is not always easy to figure out when the space will be open because it is usually not professionally staffed. Each space has different rules about who holds keys to the space and whether there are regular weekly open lab times.

I imagine the feel is similar to other hobbyist spaces. However, the equipment that fills these lab spaces is often very expensive and was collected via relationships members have with institutions or from craigslist and other listings of discarded lab equipment. Although the lab equipment is more expensive than that needed for tinkering with ham radios, it would not be considered expensive compared with traditional biotech labs. Therefore, the members likely do feel like they are making do with less (comparably). This also means that what is possible to create in the spaces is limited. In later years, some of these labs have received large grants that have helped them to establish themselves further.

For the most part, however, these spaces and groups cannot do intensive research and cannot produce a large amount of output. Nevertheless, the members know that they have been under scrutiny in the debate about whether "ordinary" citizens should have access to tinker with biological materials. Despite being small in number, the amount of discussion in government, the academy, and somewhat in the news about this kind of tinkering makes it an important site to understand what it means for changing definitions of scientific belonging, bio-economies, and biocitizenship.

In recent decades, with the explosion of biotech and its merger with academic universities, many have examined whether there is a new relationship between biology and capitalism (Helmreich 2008). Do the biological and genomic sciences of today represent a new form of capitalism? Terms such as *biocapital* and *biovalue* have been popular and arguably important for analyzing these new directions in biological sciences (e.g., Rajan 2006), but it is unclear whether the manipulation of biological materials through genomic sciences represents a particular kind of capitalism or simply fits into global capitalism in the twenty-first century (Birch and Tyfield 2013). Kean Birch and David Tyfield (2013) suggest that many of these analyses assume bioeconomies are operating as commodity exchanges without tracing the labor involved in producing value; further, they miss the fact that most of these industries do not actually produce and exchange anything but rather are based on a knowledge economy that is better represented through following assets than commodities. In this way, the academic move to take biology or "life itself" seriously may play ironically into this knowledge economy that gives "life" or "biological matter" some special status. Some have criticized these moves for assuming an intrinsic economic value of biological materials. This is not to say that a revaluation (in the sense of ethics, not capitalism) of why biological matter matters is not important. Nor is it unimportant to understand how biopolitics (Foucault) or bio-economic Man (Wynter) fits into synthetic biology and DIY biology.

Part of what is happening in synthetic biology at large and other genomic business-oriented science endeavors is an obscuring of the labor force that extracts and manipulates biological materials for use in biotech labs. Unnamed tech workers using large data processing machines process DNA-work around the clock to return results to researchers by the time they arrive in the morning to begin their higher paid work of producing publishable scientific analyses. The idea that the biological matter just appears or even self-reproduces without other forms of low-paid, exploitative labor for extraction or production of materials needed fits within a larger shift in capitalism whereby automation is promised as a future where laboring will no longer be necessary (Atanasoski and Vora 2015). In addition, intellectual ownership, direct piracy of knowledge, biological materials, and land create the scarcity necessary for the owners/financiers to benefit from the speculative work of what could be produced from genomic manipulation of biomatter.

Democratic sciences are an interesting site to understand within this larger bio-economy. Within biological sciences, the means of (knowledge) production continue to be owned by dominant classes despite various challenges by activists to "seize" these means (e.g., see Murphy 2012). More recently the widespread frustration at rising medical costs, soaring pharmaceutical profits, and increasingly blurred lines between the academy and biotech have been part of a widespread crisis in trust in biosciences. One way to deal with a lack of trust is to produce a positive affective attachment to science through methods such as science literacy.

INTRODUCTION

Another way is to claim that science is a populist endeavor without actually shifting power. And yet another way would be to change the practice of science and truly democratize the means of knowledge production to serve interests such as justice, sustainability, and equity instead of being responsive to the dictates of capital accumulation.

The knowledge produced by biosciences has been integral to how our society has operated and maintained inequalities over the last few hundred years. The value chain of knowledge production is not simply summed up by the value produced from the peer-reviewed scientific papers, which could be seen as the tangible output. Instead, this form of production has always produced truths about both raw materials and the laborers who shape them. Biology developed together with the rise in industrial capitalism in the late nineteenth century, producing what Wynter (2003) calls bioeconomic man or *Homo economicus*. The assumption of an inherent economic drive became located in our bodies through biology, and those with the appropriate/inappropriate drives conveniently mapped onto racialized and gendered bodily forms. Therefore, the struggle over who controls biological knowledge production is crucial. This is what makes bioeconomies particularly important and different than other aspects of capitalism. To understand how biological sciences impact capitalism, we must also understand that there are other things besides money that circulate to maintain our gendered racial capitalist system. *Labs of Our Own* pays particular attention to how affective (bio)economies operate.

Sara Ahmed (2004) provides us with a critical framework for thinking about the social, material, and psychic production and consequences of *affect* through circulation (affective economies) and the idea of "stickiness." She argues that affect does not reside in a body or subject but is produced through what she analogizes to the circulation of commodity and money to produce surplus value (capital), "offering . . . a theory of passion not as the drive to accumulate (whether it be value, power, or meaning), but as that which is accumulated over time" (120). Ahmed argues elsewhere that "affect is what sticks, or what sustains or preserves the connection between ideas, values, and objects" (2010, 29). In her example, "happiness" is used in maintaining the assumption of the (heteronormative) family as a social good. The way that the appropriate affective response to the family of happiness is rehearsed repeatedly creates a stronger attachment between the family and social good. To succeed in this "economy" of happiness, then, is to be happy about the family and participate in circulating more happiness through love of your family and/or the idea of the Family. Ahmed argues that when one is unhappy about the family, they are seen as the problem—the source of unhappiness for all. Applying Ahmed's work to the case of science, "passion" and "love" for science "sticks" the idea of Science as the best way to gain knowledge (epistemic authority) and sticks values of fairness, equality, and democracy to Science/Knowledge. This is the sticky affective terrain in which Science's inclusion on a list of social justice issues and marching for Science makes sense.

A positive affective orientation to science and particularly genomics in the twenty-first century is part of what has been described as a new biocitizenship.

The proper biocitizen must channel the love and trust of genomic science into participation. This participation can be found in various forms from understanding identity and health through genomic sciences to subjecting oneself to biomedical surveillance and interventions. Conversely, Ruha Benjamin (2016) names those who refuse "biodefectors." Recognizing the dangers, risks, and limitations on such individual choices, we need to think about collective responses. Can we intervene in biocapitalism and biocitizenship projects through democratic sciences?

Do Feminist Scientists Tinker?

Despite its masculinist overtones, in many ways the idea of tinkering seems like a good fit for feminists interested in engaging more directly with science. First, there is a current association of "tinkering" with more democratic forms of scientific knowledge production, which feminists have long been interested in. Second, feminists have already theorized with many of the traditional and colloquial definitions of "tinkering" as a verb (e.g., playfulness, messiness). Perhaps this is why feminists and social justice educators have begun to join in on the current enthusiasm for tinkering.[8]

The argument for tinkering versus engineering has been made through the last half century from a variety of different locations.[9] A tinkering approach as opposed to engineering approach to biology was suggested by an early article making the case for tinkering as a better metaphor for how nature evolves and therefore how scientists should engage in genetic work (Jacob 1977). From a feminist perspective, Linnda Caporael, E. Gabriella Panichkul, and Dennis Harris (1993) describe tinkering as "a particular cognitive strategy, a way of knowing, that is historically associated with low status. Tinkering is a form of adult play behavior repudiated in rationalistic culture, which holds scientific and economic thinking to be the ideal against which through and ideas are evaluated" (75). They argue that to correct for the devaluing of women and their ways of knowing, engineering fields should embrace tinkering. Although the idea of linking standpoint theories with tinkering to develop a feminist ethic of tinkering makes sense, the attempt to root it in a biological explanation—a universal "cognitive strategy"—leads us down a dangerous path. Although their argument about women's ways of knowing is more nuanced, Caporael, Panichkul, and Harris argue that infants use tinkering and therefore it is a universal way of knowing. The context of their intervention is that they are interested in diversifying engineering fields. Their suggestion and prediction are that if we revalue tinkering as a way of knowing, women's knowing would be revalued as well.

Unfortunately, their hypothesis has been proven false. *Labs of Our Own* demonstrates that that today's revaluing of tinkering has not resulted in this hopeful revaluation; instead, tinkering has resulted in similar racial and gender exclusions through its universal, colorblind, genderblind uptake as a neoliberal embrace of play and entrepreneurship. *Labs of Our Own* offers a different kind of feminist

INTRODUCTION

tinkering that relies heavily on work by María Lugones on playfulness and feminist science studies scholar Deboleena Roy's work on playful iteration of experiments. Tinkering is a method that has been and continues to be assembled and reassembled. By analyzing differences between assemblages, comparing and contrasting them, the underlying politics central to and potential consequences of each configuration become clear.

Claiming Labs of Our Own

To articulate a need for space of one's/our own implies that there is a group who has space and a group who do not have such space. The claim for such space sets up the problem and produces groups with a shared identity around exclusion. The material-discursive outcome of such a claim is not necessarily a physical space, nor is it necessarily the desired outcome. In the case of feminist scientists' calls for labs of our own in the academy, physical space is not what is most important in our calls but rather control over knowledge production and legitimacy as knowledge producers (chapters 4 and 5). And even in the case of lab space for DIY synthetic biology tinkerers in community labs, the important result is not the few physical spaces that have popped up but the defining of two sides: tinkerers versus traditional science spaces (chapters 1–3). This has implications for other critiques of traditional lab spaces, including battles over defining the bounds/purview of scientific ethics.

In the construction of labs of one's own, "lab" can be a claim for access to and/ or control of scientific methods, materials, physical space, and epistemic authority. The access and control that is sought in each of the examples in *Labs of Our Own* is over what kind of research questions can be asked and what kind of methodologies are used. The reasons for the desire for this access and control differ between articulations. Different worlds are produced, imagined, and desired by different claimants. In each case, the claim itself does work toward shaping our larger world, regardless of physical space.

Workshops and hands-on skill sharing in DIY communities create not only physical products but also community formations and affective attachments to the technical products, each other, and particular politics (Dunbar-Hester 2014; Kelty 2008). Therefore, what kinds of communities are welcomed into community lab spaces and what kinds of communities are formed by these spaces are important. In the case of DIY biology labs, colorblind, genderblind politics define the collective as "everyone" who has a desire to innovate. The political underpinning of this is based on an assumption that the natural goal is entrepreneurial success. In feminist academia, we must also be careful about how we attach affectively to science and what kind of inclusions and exclusions we might reproduce through a strong embrace of science.

Labs of Our Own comes out of my work in feminist science studies and my interest in ultimately asking how can and should we in feminist science studies imagine Labs of Our Own. Feminist science studies itself has a problem of inclusion and exclusion. This is a larger problem with academia and specifically with the two main fields that feminist science studies draws from: science and technology studies and feminist studies. Although some important strides have been made in feminist studies thanks to decades of direct interventions (Visperas, Brown, and Sexton 2016), feminist science studies seems to have retreated more often into direct collaboration with Science than with Black, Indigenous, decolonial, and queer studies that have influenced other areas of feminist studies such as feminist history, literature, and sociology. Why have feminist sciences lagged behind other areas in feminist studies? And what can happen for feminist science studies analyses if we start with insights from these fields?

To better ground feminist science studies in the political motivations of a feminism at large that seeks to dismantle systems of oppression, *Labs of Our Own* applies lessons from critical fields that are not commonly used in feminist science studies along with activist knowledges outside of the academy to suggest ways forward in producing new feminist sciences. Although biocapitalism has been of interest for those in science and technology studies and feminist science studies, there has not been a sustained focus on the intersection of race, gender, and capital (Helmreich 2008). In tracing gendered racial capitalism through democratic biology, *Labs of Our Own* asks, How does the production of a new scientific subject, *tinkerers*, rely on and shift boundaries of belonging?

Sylvia Wynter's work provides a new way to think interdisciplinarily or transdisciplinarily or perhaps antidisciplinarily by tracing the ways that the humanities and sciences have worked in tandem to produce this concept of the human (1984, 2003), and calling for feminist, Black, and other critical "area" studies scholars to work on rewriting knowledge through a new kind of science and redefinition of the human (1984, 1994b, 2003). Her direct call to academics to be accountable for the on-the-ground impact of our knowledge production for the lives and deaths of the racialized class of people deemed "not human" provides a direction and role for those of us in the academy. The stakes of our work in the production of democratic sciences must be the revaluing of life based on political direction determined by those most vulnerable and thought to be disposable in our current system.

After the police beating of Rodney King in 1991, the term "No Humans Involved" or "NHI," as used by the Los Angeles Police Department, was revealed to be in widespread use throughout the United States to refer to poor Black people, prostitutes, and others who were not considered to be deserving of basic rights. Wynter argues that academics are at least partially responsible for this exclusion in the definition of human and importantly that we as academics are responsible to work to change this definition and to follow the lead of those on the streets who are creating new definitions and new worlds through their rage and activism. The current Black Lives Matter (BLM) movement can be seen as a widespread confrontation with the kind of knowledge production that deems poor Black people nonhuman and therefore not deserving of rights. In one of the first cases that reached nationwide attention through BLM, Michael Brown was killed by police officer Darren Wilson in Ferguson, Missouri. Wilson testified that he feared for his life because to him Brown looked like a "demon" (Williams 2014). Shortly before BLM came

INTRODUCTION 19

into existence, Black activists had compiled data showing that this kind of horrific event is quite common, with Black people being killed by police or vigilantes on average at least once every 28 hours in the United States (Eisen 2014).

In an ironic move, focusing on the negative effects of Science on racialized populations may act to reinforce the same racialized differences through the repetition of traumatic knowledges over and over (McKittrick 2021). Those in feminist science studies may be part of reestablishing racialized and gendered roles as if they are deterministic—even if not naturally, socially. These repetitions may keep us stuck in colonial frameworks instead of learning from critical analyses that include constant resistance. Drawing on the concept of racial capitalism (Robinson 1983), a core concept of critical ethnic studies, *Labs of Our Own* locates examples of resistance that produce material conditions of more than simply survival but instead represent alternative forms of living and being (Melamed 2015). This is in line with a commitment to decoloniality that is interested in disrupting ideas of the individual, and of the distinction between reason and emotion in producing new ways of knowing/being (Maldonado-Torres 2016).

The question of how to create feminist sciences that are views from somewhere instead of nowhere has been central to feminist science studies (Haraway 1988). Feminist epistemologists and feminist theory texts include Donna Haraway's and Sandra Harding's work in this area as central to the field of feminist studies at large, but rarely are non–science studies scholars who think more broadly about the question of knowledge production and standpoint theories brought in for analysis in feminist science studies. For questions of democratic science, *Labs of Our Own* brings to bear insights from queer and women of color feminist scholars whose work is often claimed through the decolonial turn (e.g., Anzaldúa, Lugones, and Sandoval), feminist methodologists and science studies scholars (e.g., Roy, Lather, and Barad), and queer and feminist theorists revaluing failure and affect (e.g., Ahmed and Halberstam), alongside activist knowledges.

The reparative work of creating new truths about ourselves and our worlds as we challenge binaries derived through Western colonization is not a matter for those narrowly relegated to the sciences. Queer, anticolonial, decolonial, antiracist, disability, and feminist scholars outside the sciences proper might offer us ways forward in tinkering and playing with knowledge production. This is not a simple matter of revaluing the emotional or irrational instead of the rational; instead this scholarship promotes third ways of knowing and challenges to the binary structures at their roots.

OUTLINE OF THE BOOK

In *Labs of Our Own*, I use a mix of ethnographic methods and discursive analyses. I conducted participant observation at two DIY community biology laboratories: Genspace in Brooklyn, New York, and LA Biohackers in Los Angeles, California. I interviewed members at these two laboratories as well as participants in their workshops and meetings to understand the emergence of *the tinkerer* in mainstream science communities. I combine these ethnographic methods with close readings of newspaper articles referring to DIY science and synthetic biology and a rhetorical analysis of the websites of thirteen DIY community synthetic biology laboratories. To understand how tinkering and playing with science is conducted in self-defined feminist spaces I conducted participant observation of feminist DIY scientists and artists, and I analyze the results from my own experiments in feminist science tinkering.

Tinkering courses throughout this book. The first site of interest is DIY community biology labs where my interest in tinkering was first piqued. I work to understand how this seemingly nonscientific, imprecise concept might act to stabilize Science's authority. This leads me to focus on affective attachments which in good feminist fashion requires turning the magnifying glass around at my own community and self. Therefore, in the last half of the book I delve into how feminists participate in, resist, rearrange, reassemble, and might reimagine tinkering. The interdisciplinary (or undisciplined) methodologies and writing of this text are also meant to be an example of tinkering themselves by using concepts from various fields across university boundaries.

Chapter 1 recounts my entry into the world of DIY biology. By analyzing public communications coming from community biology labs and my own time visiting these laboratories, I attempt to answer the question of what is being produced in DIY labs. Each step of the way I describe how elusive the answer seemed to be and why: they are extremely small, their membership is inconsistent, and they are not actually producing many publishable, tangible, and/or profitable results. This leads me to ask what besides biomaterial might be produced.

Chapter 2 highlights how the production of a new kind of scientist makes further analysis of DIY biology critical for understanding science, democracy, and their intersections. I show how *the tinkerer* is produced through reliance on democratic discourses as a type of (neo)liberal subject deserving of rights. The tinkerer is formed as a specific type of person through the idea of natural types of people with different abilities. In this specific case, what makes someone belong is a desire and passion toward science. Also central to the tinkerer identity is the claim that they are excluded and therefore need space of their own. I find that the need for space of one's own is the space for entrepreneurship.

In these labs a co-optation of inclusion politics strengthens the assumption that entrepreneurial success is the ultimate goal of democratizing science. In circular logic, those with entrepreneurial desires are those who determine the terms of ethics debates and the direction of research. Truly anticapitalist drives are unintelligible. A particular scientific entrepreneurial subject, *the tinkerer*, is produced through this move to democratize science. The tinkerer is a figure of gendered racial capitalism, continuing to separate out those who belong and those who do not based on those who have the proper affective relationship to science and entrepreneurialism, thereby being overrepresented once again by white, economically privileged men. This marks a shift from using measures of intelligence to determine who belongs in science; instead, the marker of belonging in these lab sites is affective desire for science. Ultimately, I argue that what is at stake in the formation of the tinkerer as a naturalized, disenfranchised group deserving rights is (1) a further racialized and gendered exclusion from the means of scientific knowledge-making through a colorblind and genderblind politics and (2) a definition of democratic science that leaves intact participation in capitalism as the natural and most desirable goal of inclusion.

Chapter 3 examines the impact of the tinkerer on current ethics debates. The tinkerer becomes the champion of the fight for a return to a more pure and innocent science that is fabled to have existed before Big Bio. The ethical consequence of this is that the tinkerer is positioned on the side of justice and ethics debates remain attached to this one narrative.

On DIY and community-based synthetic biology organizations' websites, democratic science is presented as a novel, progressive approach to science that addresses ethical concerns and at the same time produces better scientific results. In the physical lab spaces, what seem to be well-rehearsed ethics discussions come up during class sessions. I argue that the rhetoric of democratic science in synthetic biology realigns boundaries between the Public and Science, whereby scientists lay claim to solidarity with the public at large in opposition to traditional biosciences and Big Bio. The superficial use of the language of rights and democracy, however, relegitimizes the primacy of scientific discovery to solve societal problems. Further, by becoming "the informed public," ethical challenges from publics critical of genetic sciences may become delegitimized. The animating question is not whether more democratic science is good or bad but rather what are the material-discursive effects of claims to a more democratic science.

In chapters 4 and 5, I turn to look at my own field of feminist science studies, focusing on trained scientists who have defected to women's studies spaces, those of us who have "failed" at science proper by leaving the science pipeline. I draw on the work of other scientifically trained feminist academics such as Lynda Birke, Ruth Hubbard, Deboleena Roy, and Banu Subramaniam to situate my analysis.

I read our desires to do science and fears of science through the insights of feminists who have been trained as scientists, who have had the "passion" to do science and "failed" at their science careers. Drawing on analyses of failure (Halberstam 2011) and willfulness (Ahmed 2014), I suggest a re-evaluation of "success" and "failure" in doing "real" science. Desire for proximity to or success at doing science must be read within the context of the automatic/unearned epistemic power that Science holds. After making the case for a careful reading of our affective attachments to s/Sciences in chapter 4, I move on in chapter 5 to focus on what I call queer sciences of failure (drawing on Jack Halberstam). I suggest that fear of or failure at Science proper might be read as an act of resistance. This analysis frames one of my central contentions: that we (feminist scholars) are in fact already doing science through our experiments in feminist studies classrooms. By claiming this space for scientific knowledge production, we might challenge the definitions of science and scientist. I use the notion of "failure" to propose a role for feminist scientist defectors in the *redistribution of epistemological authority.*

These chapters focus on how "labs" and "our" are defined by feminist scientists when we express our own longings for "labs of our own," imploring us to be careful to not fall into a trap of simply claiming more space for us as scientifically trained scientists.

In chapter 6, I use feminist and decolonial theories to define feminist tinkering as an explicitly politically motivated "playing" with the epistemic power of Science. How might we draw on feminist theories to find ourselves tinkering as a challenge to the status quo? I bring together examples from feminist DIY community science labs, previous feminist and social justice movements, and feminist classrooms to theorize tinkering as feminist praxis, tracing an alternative genealogy for today's democratic DIY sciences from activist movements of the 1970s and 1980s. Once again, I focus on the particular framings and justifications for needing "our" own space—looking at how definitions of "our" are framed politically. I include cyberfeminist art collective subRosa; the now defunct reproductive and environmental justice organization Committee on Women, Population, and the Environment; and Barcelona-based biohackers and post-porn activists and artists oncogrrrls and Quimera Rosa. I analyze examples from my own attempts at feminist science in the classroom that bridge academic boundaries through collaboration with community activists. In arguing for a more complex, messy feminist method, I draw on Maria Lugones's (1987) arguments for embracing "playfulness," along with Deboleena Roy's (2012) embodied and situated approach of "feeling around." Ultimately, I suggest that "tinkering" can be used to put these feminist theoretical contributions into practice inside and outside traditional science classrooms. Using Chela Sandoval's theorization on decolonial love (2000), I come back to the relationship between affect, community formation, and ethics in the development of a feminist approach to tinkering with science.

Labs of Our Own argues that claims for and attempts at democratic sciences not only impact what counts as science and who counts as a scientist but reconfigures who is included in the demos. Because of this redefinition of publics, uncritical excitement at more democratic forms of science is dangerous. Instead, I call for a critical analysis of each site, examining the political and ethical consequences of how affective claims for inclusion and exclusion in science for "everyone" or "us" are made. *Labs of Our Own* builds a case for a feminist, antiracist, decolonial, queer science tinkering practice that intentionally, politically, and ethically acts to produce new kinds of sciences, bodies, and cultures.

INTERLUDE 1

Serendipity

One morning I opened my email (which I was not supposed to do before my morning writing) to find that Genspace, one of the labs that I had been following online and through ethnographic fieldwork, was announcing a move from their downtown Brooklyn location to Sunset Park, another neighborhood in Brooklyn. The email contained an invitation to attend their "housewarming" party at the new location—a little over a mile from where I grew up and where much of my family still lives. The move to what I still call "my neighborhood" made this project suddenly feel more personal. It made me remember how I learned the words "serendipity" and "gentrification" on the same day, and it made me wonder what (if anything) this move meant for the current politics of gentrification.

From what I remember, I was about ten or eleven years old (in the late 1980s/early 1990s) when my dad stopped the car on the way home only a few blocks away from our house. He pulled over to what looked like abandoned lots. There were fences around them, and a sign on one fence said "Serendipity." The location was a different direction from my junior high school and friends' houses, so I had not noticed these changes happening to the block. But the adults in the neighborhood were very aware of what had happened. Over time, nearly the whole block of houses had burned down in a series of fires that took a few homes at a time.

I remember my dad asking me if I knew what the word *serendipity* meant. He explained it, and he told me what most in the area suspected (or knew?): the fires had been set by hired arsonists for a company that was benefiting from the properties' destruction.[1] What strikes me now is I already knew at that time that this was not a farfetched idea. While it was clearly wrong and something to fight against, I do not remember it being overly surprising that a group could commit arson and then buy up the land inexpensively to begin the process of what I also learned that day was called *gentrification*.[2] The kicker was that they had the nerve to name themselves Serendipity.[3]

What is the relationship between serendipity and tinkering? The myth of serendipity that Serendipity Associates, Ltd., the real estate company, represented is

23

what brings this story together with tinkering, scientific discovery, and innovation. Scientific discovery has a long history/relationship with colonization. The *Oxford English Dictionary* (OED) definition of serendipity, "the faculty of making happy and unexpected discoveries by accident. Also, the fact or an instance of such a discovery," includes as one of its examples Columbus finding America instead of "the Indies."[4] The common use and definition of serendipity refers to something unexpectedly happening that is beneficial to the parties experiencing serendipity, so serendipity is often thought of as a lucky occurrence.

However, business professionals in the *Harvard Business Review* have argued that serendipity is not about luck but about creative discovery (De Rond, Moorhouse, and Rogan 2011). That is, you cannot just sit around and hope that something serendipitous happens, you must actively try to make things work—only then will you be in a position to receive serendipitous success. They include tinkering specifically as one way to embrace serendipity—to discover something that you did not set out originally to discover.

As we see in the OED's example of serendipity to describe colonial discovery, the problem is that what one person considers serendipitous, another may give a very different meaning. In that example, of course, the subsequent genocide was clearly not a serendipitous event for those murdered. Film director Spike Lee drew together the relationship between the supposed discoveries of colonizers with current day discoveries of gentrifiers when he discussed the case of Fort Greene, Brooklyn: "Then comes the motherfuckin' Christopher Columbus Syndrome. You can't discover this! We been here. You just can't come and bogart" (Michael and Bramley 2014). The anger exhibited by Lee is not uncommon from those who are not experiencing the serendipitous effects of gentrification. In the example I began with, the community I grew up in has continued to push back against gentrification. The serendipity of the real estate company to make money was not serendipitous to those living in the buildings before the fires or to the broader community that is being displaced through ongoing gentrification.

The idea of serendipity is linked to the idea of pure objectivity in science and its myth through this belief in discovery that values the discoverer's perseverance and creativity but does not consider how the discoverer's cultural and political location impacts how we understand what is discovered. This omission of the positionality of the discoverer in the definition of what is discovered means that other ways of defining this truth are erased. This idea, however, is strongly linked to the idea of a pure objectivity. While crediting the ingenuity of the scientist for finding the way to that truth, the subjectivity of the scientist and other cultural factors are generally removed from how we understand the discovery.

Feminist science studies scholars and others have critiqued this understanding of isolated, universal truths. The belief in objectivity leads to a belief that if you just keep trying enough things you will be able to discover the answer. Here lies the myth that anyone can do it with some work and luck! The myth of objectivity therefore is related to the myth of serendipity that I have described. In the first chapters, I trace how the scientist's role in discovery is changed from the modest

INTERLUDE 1

witness to the tinkerer without challenging the idea of objectivity—perhaps even strengthening it. Part of what is behind tinkering and a push for more democratic science is not fairness as much as having more people tinkering, more people exploring, and therefore more people discovering. This assumption and belief that truths out there are waiting to be discovered means that there is no interest in how the person doing the tinkering matters or in how the political and cultural conditions of the tinkering (the purpose) matter. So even though the belief in objectivity means that the truth is assumed to be out there waiting to be discovered, what I relate in *Labs of Our Own* is a story of how the discoveries and discoverers in my story have been overdetermined.

What about biohacking and gentrification? Genspace is not alone in finding its home in gentrifying neighborhoods. When I visited LA Biohackers in Los Angeles, I found myself driving just a little bit past Skid Row to an area with some old warehouses that were being used for various smaller enterprises.[5] In the same building where LA Biohackers operated, there had been major drug busts, with police SWAT teams descending on the building (according to the biohackers). During one of my visits our research team met individuals in the elevator who were going to shoot a film inside that, when we inquired, seemed to be pornography. Just a few blocks away, however, we could find a coffee shop with soymilk. This led us to believe that the area was on the cusp of gentrification; a lab member told us that the area was "in transition," which seemed to verify our assumption. Similarly, when I visited the Baltimore Underground Science Space (BUGSS), I found it in a complex that was clearly part of the hopeful gentrification in Highlandtown, as evidenced by the security bars over their windows obscuring their logo. Highlandtown is known to be a gentrifying area in Baltimore, which has been leading to decreased racial diversity (Filomeno 2017).

Although these community biology lab spaces are small and do not have much power to take over large amounts of space, the impact of art spaces on neighborhoods has caused some activists to be similarly concerned about the use of science in processes of gentrification (Free Radicals 2017). As we will see in the next chapters, these labs are part of the similar ethos of business incubation that created companies like Google in the tech world. One of the most extreme cases of gentrification has taken place in the San Francisco Bay Area where technology companies have been blamed for the influx of wealth that has displaced poor and middle-class people (Rosalsky 2021). This industry has created its own privatized services, such as transportation (Gentile 2021), thereby not investing in the existing public systems while making it impossible for even middle-class professionals to afford to live in the area. The number of unhoused people in the Bay Area has been a topic of wide concern, with numerous legislative, charitable, and activist interventions aimed at addressing the problem.[6]

In New York City, Sunset Park's gentrification has been somewhat slower than others, only ramping up in the last decade. There has been active resistance over this time (as with most gentrifying areas), and with it have been both discursive and material fights to change the conversation. What is different for me is that I know this story more intimately and have followed it more closely. Does it matter at this point that Genspace is moving into the neighborhood? Obviously, DIY biology is not the reason for the displacement of hundreds of thousands of people and the rebranding of the neighborhood. But I do think it is important in thinking about the claims that these labs make to social justice to consider their relationship to their new "home."

As I drafted this interlude after receiving the email, I wondered who would be at Genspace's housewarming. Perhaps the lab would become an integrated part of the community and a community hub for resisting ongoing displacements. The kinds of communities that are formed through different democratic science formations is central to this book. Through the analysis in the next chapters, I map out possibilities and dangers in moving forward toward more democratic sciences.

CHAPTER 1

(De)constructing DIY Community Biology Labs

In the early 2000s, synthetic biology and do-it-yourself (DIY) biology began to take root, cross pollinating each other along the way (Calvert 2013; Meyer and Vergnaud 2020). Synthetic biology (also called SB or syn bio) promised to bring engineering methods to biological matter to create new, more controllable ways of manipulating nature's systems. Some saw synthetic biology as a rebranding of genetic engineering, at that time widely perceived as an ethical landmine. Although this may be part of the story, synthetic biology also brought new collaborations and methods together creating new communities beyond biology proper. In 2008, at the fourth international synthetic biology conference in Hong Kong, a handful of DIY biologists found each other among the other conference-goers and thus began an international network of DIYers. Shortly after the conference, Jason Bobe and Mac Cowell founded a message board (and eventually website) under the name DIYbio; it began with about twenty members and had grown to thousands by 2015 (Meyer and Vergnaud 2020).

A few years after the forming of the DIYbio network, the Federal Bureau of Investigation invited several DIY biologists to New York City, all expenses paid. Cameron, one of our DIY biology informants who I interviewed in 2015 and 2016, was among them. "We showed up on the street corner where they told us to meet. A black van with darkened windows pulled up. Inside they took our cellphones.... The van was parked in an underground parking garage, [and] we met in a windowless room. We didn't know where we were taken" (Cameron, interview, March 22, 2015).

My collaborator and I were surprised when Cameron recounted these events. Cameron did not seem overly concerned as he described what he called an "abduction" by the FBI. We wondered, Were they scared? Why did they get into the van? Cameron believed the FBI did not realize how "creepy" they seemed doing this—he described the FBI as "clueless" at the time because there was nothing to be concerned about. Cameron's sense was that after meeting with the DIYers, the FBI understood that they did not pose a threat. I, on the other hand, did not think there was any mistake made with the tactics chosen by the FBI agents—intimidation and producing fear are intentional.

My collaborator and I are both activists. Stories of surveillance and intimidation by the FBI, Homeland Security, and CIA are not foreign to us. Activist movements with goals to create radically (and even not so radically) new societies are well-accustomed to security/policing arms of the state moving in quickly to defend any potential attack on global capitalism. Did the FBI's interest signal that DIYbio constituted such a threat?

Over the last two decades much has been written about whether synthetic biology and DIY biology represent a major biological shift. Sometimes DIY biology groups have explicitly claimed (scientific or medical) revolutionary aims. I became interested not only in what was happening in these spaces as far as scientific experiments but in what kind of political communities were being produced. DIY biology communities aimed to create a more democratic way of doing science open to all, and some DIY biologists were able to create physical spaces open to the community where hands-on biology experiments and community building could regularly take place. What did the formation of these communities mean for capitalism, nation states, and the social justice movements that critiqued them?

In December 2010, Genspace opened in Brooklyn, New York, claiming to be the first community biotechnology lab in the world. By July 2014, the Synthetic Biology Project (a project of the Woodrow Wilson Center) identified on its website (www.synbioproject.org) thirteen community lab spaces in cities in eight countries around the world, including Australia, the Netherlands, and the United States.[1] These physical community labs were preceded by the 2008 founding of DIYbio as an international network of DIY biologists emerging from the larger synthetic biology community. In January 2016, according to the DIYbio.org website, there were seventy-two self-identified DIYbio groups worldwide, some with dedicated lab spaces and others using meetups virtually and in-person to support each other in their kitchen and garage biology projects (Keulartz and van den Belt 2016).[2]

DIY Biology's Websites

Between 2014 and 2016, I visited physical DIY laboratory spaces, participated in classes and open lab activities, interviewed members and participants, and followed news coverage and policy discussions, at times with the help of research assistants. Before visiting any of the labs, I looked over their websites to get a sense of the public image and information the community biology labs put forth. The DIY, hands-on component of their work was prominent, as was highlighted on the original website for Baltimore Underground Science Space (BUGSS): "BUGSS emphasises [sic] understanding through DOING; a space where it is ok not to know."[3] This "DOING" was often in the interest of better learning or education: "understanding through DOING." Community biology labs were offering the opportunity for everyday people to come in and learn, whether or not they had any experience: it was "ok not to know." Genspace similarly highlighted that they "offer[ed] biotech classes to people with no prior lab training."[4]

The claim that experiential, hands-on learning was better than traditional education was directly stated on Genspace's website: "Taught by members with doctoral level expertise, we provide an immersive experience that no lecture course can match." This idea was exciting to me as someone who worries about the gatekeeping nature of the academy and values popular education. Further, the community biology labs offered access to training on real equipment with real biological materials. By providing these resources to the amateur, the promise was that the general public would be able explore and experiment as they wished.

The websites of BOSSLab (Boston's Open-Source Science Center) and LA Biohackers explained how this information was shared: "The mission of the Los Angeles Biohackers is to make science accessible to people of all ages and educational backgrounds. We do this by providing laboratory equipment and workspace, resources which are typically out of the reach of the amateur, to anyone passionate about learning," and "The mission of BossLab is to open scientific exploration to all & to help solve problems we face on Earth."[5] BOSSLab's vision explicitly linked democratizing scientific exploration to solving real-world problems; although other labs' sites were less explicit, the argument that more exploration would lead to societal benefits was common.

The method of exploration was repeatedly tied to entrepreneurship in laboratories' self-descriptions. In the self-defining statement on the BioCurious site, entrepreneurs led the list of people the space was designed to support, and education/training was also emphasized:

BioCurious is . . .
- a complete working laboratory and technical library
- for entrepreneurs to cheaply access equipment, materials, and co-working space
- a training center for biotechniques, with an emphasis on safety
- a meeting place for citizen scientists, hobbyists, activists, and students[6]

The BioCurious statement contains another element that appeared on many of the websites: they offer a place to form community and share ideas toward a more democratic vision of science.

Genspace articulated another common thread, that one of the things one could *do* in the space was to enjoy science, to have fun: "Genspace is a nonprofit organization dedicated to promoting citizen science and access to biotechnology. . . . As a community-based lab, we offer members the unique opportunity to work on their own projects and experience the joy and wonder of science firsthand." The fun to be had is a theme for not only those new to science but those who labor daily for science under the constraints of biotech or the university.

The idea of these labs being places where it is "ok to not know" references an opening up of science not only to those not (yet/formally) trained in the sciences but also to the idea of tinkering as a way of doing science. *Tinkering* has often had negative associations related to messing around with something without knowing what you are doing. In many ways tinkering seems to be oppositional to the way science, a disciplined field of learning that follows the scientific method, is supposed to be conducted. Yet tinkering encompasses the joy and fun and discovery promised by these spaces. Tinkering signals a space for the amateur. Amateurish experimentation also suggests that the end goals are less prescribed and the sense of urgency and structure to the experiments might operate outside of the time frame of the paid sciences.

The prerequisite to join in DIY biology was not any particular academic degree, or standardized test, or previous coursework, or proof of success at math or science. These measures have often been used as a supposedly neutral way to discriminate against and exclude people of color and women. Instead, according to community synthetic biology websites, democratic-oriented synthetic biology projects are for "everyone," "all of us," "anyone," "the public." Given my deep interest in more democratic sciences, these labs, in theory, were perfect for me. I certainly felt welcome. But were these spaces that I felt called to become a real part of? Who was truly included in community DIY labs?

These organizations were looking for the type of person who wants to "tinker," "explore," and "create." This may include those who identify as tinkerers, innovators, inventors, hackers, artists, scientists, explorers, and or entrepreneurs (Table 1.1). Passion for science appears to be the most important marker for inclusion in these community biology laboratories. As the Biospace site suggested, if you "like to tinker and learn" these labs are perfect for you. There is a similar emphasis on a desire to be there on many of these sites. BioCurious uses the language of "anyone who wants to experiment with friends"; Biologigaragen in Copenhagen said, "anyone with an interest in the natural sciences"; and LA Biohackers used the phrase "someone who enjoys" to describe the kind of hackers they are.

Through keywords such as "want," "enjoy," "like to," and "interest," a call emerges for those excited about biology to join. As La Paillasse in Paris puts it, community biology labs are "group[s] of *passionate* people about biology" (italics added).[7] This passion supersedes an innate ability to do science, thereby opening up the possibility that many more people can "create" and "tinker" together. There is one outlier in this messaging; LA Biohackers uses the word "aptitude," which does suggest an innate ability. It is unclear from this one usage, which seems to be out of place among the others, whether it signals a deeper, continued belief in some people's natural ability to do science better than others. In context, it seems to be a way to signal that they want people who are serious about really doing experiments and learning rather than pursuing some of the aesthetic aspects of other hacker communities, which they seem to disdain. That kind of seriousness also fits into the broader emphasis on passion for science.

Although the prerequisite of certain schooling and/or natural intelligence was almost entirely absent, a feeling toward science—a passion to tinker—does

(DE)CONSTRUCTING DIY BIOLOGY LABS

TABLE 1.1

WHO BELONGS? WHO IS PART OF THE COMMUNITY? QUOTES FROM
COMMUNITY BIOLOGY LAB WEBSITES THAT DESCRIBE THE TYPE(S)
OF PEOPLE WHO ARE WELCOME AT THE LAB.

- A Nursery for *Explorers* and *Entrepreneurs*. (Genspace)

- We bring together *engineers, scientists, attorneys, innovators, teachers, students, policymakers*, and *ordinary citizens* to make this vision a reality. (Biobricks)

- *If you like to tinker and learn* more in the area of biology and its intersection with other technologies, this is the place for you. (Biospace)

- Designed to facilitate more *independent researchers* and science *entrepreneurs* to pursue their work full time. (Brightwork)

- We're building a community biology lab for *amateurs, inventors, entrepreneurs,* and *anyone who wants to experiment with friends*. (BioCurious)

- For *entrepreneurs* to cheaply access equipment, materials, and co-working space. (BioCurious)

- A meeting place for *citizen scientists, hobbyists, activists*, and *students*. (BioCurious)

- To us, a *hacker* is someone who enjoys solving problems, taking things apart, building things from scratch and prioritizes technical competence and aptitude over appearances or hubris. (LA Biohackers)

- BUGSS is a laboratory space for use by *amateur, professional*, and *citizen scientists*. (BUGSS)

- Biologigaragen is a laboratory and open creative space where *anyone with an interest in the natural sciences* can meet, play and share their ideas, thoughts and inspiration. (Biologigaragen)

- BioArt Laboratories (BL12) is a place where *artists, designers* and *students* meet, a place that can offer the ability to explore developments in the field of life sciences in an innovative and creative way and where experimental work can be done. (Bioart)

- The Bio-hackerspace La Paillasse is a group of *passionate people about biology,* each with his or her own area of expertise, interest and dedication. Some are from the *computer* side (read *geeks*), some others are *PhDs, researchers, professionals* and *students from the biology* side and some others are *designers, artists, engineers*, all of them want to see new exciting concepts and create interesting projects using biology as a primer. (La Paillasse)

Note: Specific terms italicized by author for emphasis.

emerge as another prerequisite. But does replacing the criterion of natural intelligence with that of passion really challenge patterns of who is included?

IN NEED OF SPACE OF THEIR OWN

Revolution as a term and idea has various meanings and political affiliations. The kind of revolution that DIY biologists invoke borrows imagery and rhetoric from more radical leftist circles (Tocchetti 2014). One prominent example is what has become the most prolific DIYbio image (Figure 1.1). On the DIYbio website, when you scroll over the image of a hand pipetting you find it labeled "diybio revolution fist." This image most likely makes one think of the Black Power and feminist movements that popularized raised-fist images and actions. The raised fist has been used for over a century for various socialist, communist, anarchist, and other leftist and antifascist movements around the world (Korff 1992). Although right-wingers, leftists, and centrists have all used "revolution," this image draws on a specific tradition.

In claiming what community biology lab spaces are and who they are for, DIY biologists also make implicit and explicit claims on their websites about what they are not and who is excluded. Implicit in the claim that they are creating a democratic, transparent, and accessible space in which to do science is the assumption that this is in opposition to the status quo of science. At times this status quo is explicitly called out, as on the Genspace site: "Unlike traditional institutions, our diversity is our strength and the source of our innovation." Genspace argues that traditional science (university and big biotechnology science) is limiting—that democratic sciences, which include a diverse group of actors, make them not only different but better at doing science than traditional sciences. Implicit in this statement but more explicit in other statements by Genspace and other labs is that traditional science keeps the means to do science out of the hands of the public:

> Access to the source of these [scientific] developments, the laboratory, is rare, and rules and protocols resulting from the scientific nature of these places, can hinder the freedom of the individual. So, a different approach is required . . . (BioArt)

> Access to knowledge shouldn't be limited to academia and all of the restrictions associated with it. (Biospace)[8]

These claims are well-founded. Critical scholars and activists have long pointed to the elite, exclusionary nature of traditional science spaces. The resulting exclusions can be used in a circular logic to show that women and people of color are not generally capable of doing science. Although the academy continues to claim space as a not-for-profit place of learning and discovery, even public universities have become more and more corporate in their practices over the last several decades (Newfield 2008). Further, at the intersections of science, engineering, and technology, the academy and technology corporations have been closely entwined in practices of sharing resources, ideas, personnel, and sometimes even profits (Mody 2006).

Therefore, the exclusionary practices of the academic and corporate worlds are also shared, hence the academy and biotech are collapsed as "traditional" biology in the counterdiscourse of community DIY biology labs. Members of community biology labs position themselves and their laboratory spaces as an answer to this problem, a way to rectify this disenfranchisement.

Who, then, is gaining greater access to science through community lab spaces? Many of the founders and participants in these spaces are formally trained in the sciences and often continue to have day jobs in traditional science labs (Grushkin,

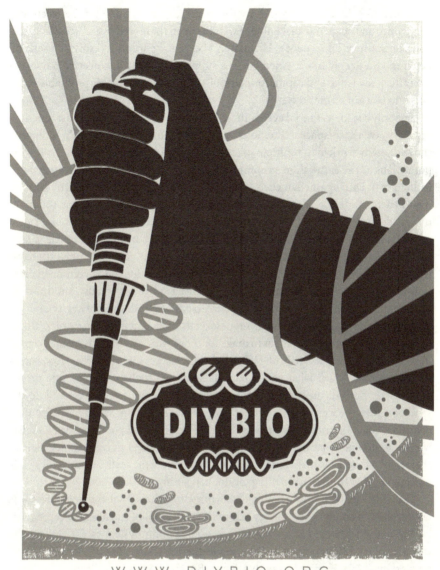

Figure 1.1. DIYbio "revolution" poster and logo. The image is a human hand using a pipette. In the color version, there is a yellow background with a grayscale fist and pipette. There are DNA strands in the background and circling the wrist of the pipetting hand. Image credit to diybio.org. Image licensed under a Creative Commons Attribution-ShareAlike 3.0 Unported License.

Kuiken, and Millet 2013). Therefore, the same co-production seen between tech companies and academic disciplines exists between these traditional science spaces and community DIY spaces. Ultimately this speaks to an irresolvable tension between two deeply intertwined goals: a supposedly neutral, objective pursuit of scientific knowledge and capitalist interest in producing patentable profitable products and consumer markets.

Not only are ideas shared across these boundaries but the same actors are moving between these spaces (Ikemoto 2017). As a 2013 survey of DIY biologists showed, many worked in multiple spaces including academic, government, or corporate labs at the same time as community biology labs (Grushkin, Kuiken, and Millet 2013). During one lab class I took at a community lab, the instructor took me to their academic laboratory to finish up the experiment and use the more powerful equipment they had access to there. Among the DIY biologists I met were graduate students, former and current professors, and corporate scientists.

It is becoming harder and harder to argue that the not-for-profit university system represents academic purity over biotech companies due to the commercialization of the sciences in and outside the hallowed university walls (Smith 2014). Community biology labs emerge as a counterspace to both the university and Big Bio, which bases its claims of deservingness on its exclusion from the former. This is in some ways a strange argument to make for those who have already been accepted in those worlds and have willfully left or in some cases remain in both. However, as someone who also left what would have likely been a higher paying, more prestigious career path in the proper sciences because I did not feel like I could do the kind of research I wanted to, I did not necessarily find the position oxymoronic. I continued to be curious about what kind of science the community biologists wanted to do and what they were actually able to complete in community biology spaces. What were their dreams and goals for open, democratic, community biology research? What was happening (or not happening) in these labs?

The collection of website statements led me to believe that the *doing* that was taking place in community biology laboratories was a lot of learning, entrepreneurial exploration, sharing of resources (knowledge and physical tools) beyond the confines of traditional science, tinkering in the interest of making the world better, and just plain having fun with science. But these are generalities. What kind of biological experimentation was actually happening in these spaces? What did they discover? How were they dealing with issues of patents and intellectual property rights? How did they deal with community conflict and differences of opinion on what should be happening in the labs?

The most advanced project that we had heard about was that of the glowing plant which became notorious inside and outside the DIY biology community. In 2013, Omri Amirav-Drory, Antony Evans, and Kyle Taylor raised almost half a million dollars through crowdfunding on Kickstarter to fund their glowing plant project. This synthetic biology project, undertaken in the BioCurious lab, brought quick fame, interest, and ultimately condemnation to the project, lab, and broader DIY biology community. The project aimed to incorporate bioluminescent-causing genes from fireflies and bacteria into the plant *Arabidopsis*. The promise, which ended up being an unfulfilled dream, was to create a biological source of light to replace modern electricity.

The team began their Kickstarter video with the provocation "What if we used trees to light our streets instead of electric streetlamps?"[9] The video splices the original recording of their pitch with that recording being played on a computer monitor in what is presumably their DIY lab space. Toward the end of the short video (2:07 minutes), a shot from the 2009 movie *Avatar*, in which the trees provide light, appears on the computer monitor, standing in for the promise of the glowing tree. On the monitor, *Avatar*'s main character looks around with wonder-filled eyes at the glowing trees and other flora. These images and storyline give the viewer the sense that this is really possible because we feel like we are in the physical lab space. To the right of the frame, we can see part of a picture of George Church, an outspoken proponent of genomics and one of the leaders of synthetic biology over the last couple of decades.

The project leaders posited the experiment as an environmental sustainability project that could radically change our dependence on nonrenewable resources, including the materials that go into creating light bulbs themselves. Environmentalists concerned with genetically modified organisms (GMOs) and the unknown consequences of this genetic manipulation critiqued the project. Most contentious was the distribution of seeds to funders, which some ethicists, activists, and policymakers moved in quickly to limit. Kickstarter eventually banned the group from giving away the genetically modified plants (Geere 2013).

Besides the ethical limits on their project, the DIY scientists found themselves unable to meet their goal because it was a lot harder than they thought to have a plant produce a significant amount of light. One of the members, Evans, reportedly told the *Atlantic* that the project was "premature," leaving open the belief that it is possible but more scientific progress would be necessary before it could become a reality (Zhang 2017).

INSIDE DIY COMMUNITY BIOLOGY LABS

The dream of the glowing plant was never fulfilled, but the project prompted me to wonder what potentially revolutionary projects people were working on in these community spaces. I began visiting the spaces to get a better sense of what happened there, what was possible, who was there, and why were they there. Over a two-year period, I visited LA Biohackers (which later changed their name to LAB) in Los Angeles, California, and Genspace in Brooklyn, New York, several times. Each one was located at the time in an old warehouse. To reach the lab spaces, I walked through poorly kept entrances (e.g., see the picture of Genspace elevator in Figure 1.2) and entered old elevators to go up to higher floors. The warehouses were home to various other low-budget projects.

One day, on our way to LA Biohackers, my research assistant and I found ourselves making small talk in the elevator with a group carrying film equipment.

Figure 1.2. Elevator door on ground floor in Metropolitan Exchange Bank building where Genspace was located. The door was painted red with lots of peeling paint spots and dark stains. The elevator has a door that opens before you enter the actual elevator, which is clearly not from this century. Photo by Sig/Sara Giordano.

We asked what they were filming to which they responded, "A reality TV show." "Cool, about what?" we asked. They answered with one word, "Court." Later that day, talking with Cameron, we found out more about what the warehouse was used for when we asked if the other renters or their landlord cared what they did in the lab:

> No. There used to be like a bunch of drug manufacturing operations in here. When we first moved in the unit next to us was making pot, like growing pot. Um, I was cool with it at first because I thought it was just some hippie kids growing some weed to sell to some dispensaries. But then it turned out it was MS-13, which is like a gang operation, and they got raided by the SWAT team, and they had a whole arsenal in there.... Yeah, the landlord doesn't care what happens in here, and there's like some porn studios ... yeah, occasionally you'll see someone walk by here with nothing on.

In addition to the possible pornographers we met on our elevator ride, we saw some scantily clad folks run by as we used a seemingly quiet space out in the hallway to interview lab members. Cameron appeared mostly nonjudgmental and unfazed by the semilegal and illegal activities that took place. The gang activity unnerved the group a bit, but the neighbor they found to be most irritating was the bagpipe studio that provided space for novice players to practice (badly). In Brooklyn, the warehouse where Genspace worked seemed to be occupied by more reputable enterprises, artists, and (legal) entrepreneurs, but the aesthetics of the building and the crowded lab space were similar (Figure 1.3).[10]

As someone trained in the sciences but no longer a practicing scientist, I entered the space ready to get my own joyous hand-on experience with genetics. I succeeded in using a pipette for the first time. My classmates and I took pictures of each other so that we could text our friends this impressive sight: I finally looked like a recognizable scientist with that pipetting hand in action (see Figure 1.4 for evidence)!

Figure 1.3. Genspace lab in a cluttered warehouse space. Photo by Sig/Sara Giordano.

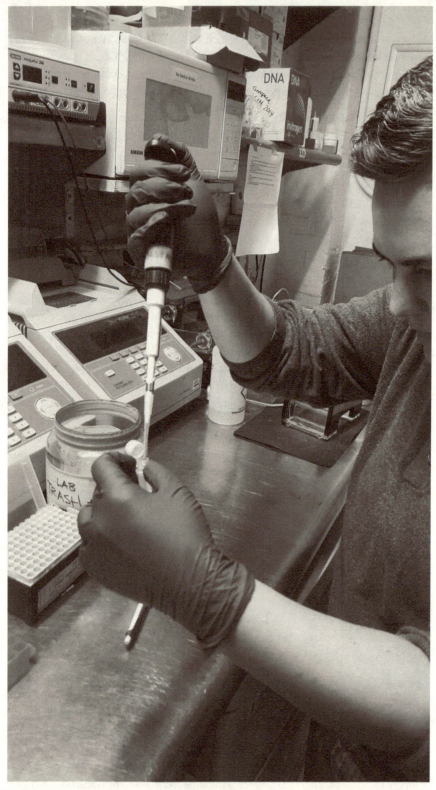

Figure 1.4. The author using a pipette during a Genspace workshop. Photo credit Ananda Gabo.

I also quickly remembered the timers that my PhD cohort members would carry around with them when they were "running gels" and similar processes that may sound exciting but in reality are slow and a practice in patience.

Before I could do any "real" experiment, of course, I needed to learn more about the techniques and the philosophy behind genetic manipulation. Genspace had two different multisession courses running. I signed up for both "Biohacker Boot Camp" and "Intro to Synthetic Biology." A third course called "Biotech Crash Course" was also offered but was the same material as the "Biohacker Boot Camp" spaced out across a longer time period so I did not take that one. Later they also added a "Genome Editing with CRISPR-Cas9" course with a prerequisite of their CRISPR workshop. I was able to attend the CRISPR workshop but did not take the full lab intensive course.

Even though I had already earned a PhD in a "real" science, this was a different kind of lab and material than I had worked with, and being in these lab spaces felt different to me. I was both nervous and excited, similar to how I remember feeling when using science kits to experiment at home as a kid. There was a novelty to the experience and a feeling of freedom to try new things for fun—as promised by the websites I had spent much time looking over before ever visiting the labs. In the classes, I did not feel the pressures I was used to in typical classroom settings; the DIY classes were not graded, of course, and the goal was for everyone to succeed, so there was a good sense of collaboration. The website promises had come true!

FOR THE LOVE OF SCIENCE

The DIY laboratories that I visited emphasized the ability to follow one's heart in scientific exploration. Dedication and love for science were ever-present. This flew in the face of the accepted image of the scientist, which for well over a century has been one of a rational man (coded as white, propertied, and male), devoid of emotional responses distracting him from objective work. Passion signals a strong and sometimes uncontrollable emotion—here, an emotional attachment to science was not only acceptable but perhaps required.

During one class that I took at Genspace, the instructor recounted a popular origin story about the creation of PCR (polymerase chain reaction) technology. PCR is used to amplify small amounts of DNA thereby making it possible to perform many analyses without ruining the only sample you have. PCR is generally hailed as one of the most important advances in processing DNA over the last century. The tale goes something like this: Kary Mullis (the inventor of PCR) was supposed to be going on a romantic vacation up the coast of California with his girlfriend but could not take his mind off science. He was so obsessed with scientific discovery that he did not pay any attention to his girlfriend. This led to the invention of PCR and also the end of that relationship. The story seemed intended to convey the amazing obsession that he had with figuring out the scientific problem. The origin story of PCR depicts a scientist as both detached from normal social understandings, evidenced in the story by Mullis's ability to maintain his romantic relationship, and also overly emotional and passionate toward science. Mullis himself had shared a version of the tale during his Nobel Prize speech and many times in interviews and in his own writing throughout his life.[11]

In Mullis's story, his love for science is linked with his failure at heterosexual love. However, this narrative does not necessarily disrupt the normative script but instead indicates the appropriate gendered roles in heterosexual relationships, where the scientist's wife or girlfriend should learn to wait for a man who is busy working on important (nondomestic) things. That is, women are supposed to be supportive of their men's scientific work, which presumably serves a higher purpose (Traweek 2009). As Sharon Traweek (2009) shows in physics, learning these gendered scripts is part of the educational process that makes someone a scientist.

The juxtaposition of science and women as different objects of love in competition for a man's attention reinforces that the proper scientist must be male. Donna Riley (2017) uses a similar example of an engineer describing his passion for slide rules as greater than his love for his ex-wife to assert white male heterosexuality as the norm in engineering cultures. Whiteness is signaled in nerd culture by being just the right amount of masculine. According to Ron Eglash (2002), the figure of the Asian man is feminized, and the Black man is too cool and hypermasculinized, thereby carving out a path for the white, heterosexual, male nerd in the middle to be the proper scientist.

In the Genspace instructor's version of the story, the possibility that Mullis's girlfriend had anything to do with how Mullis thought about science or the world was removed. However, in his Nobel Prize speech, Mullis notes that his girlfriend was a chemist who worked at the same company where he worked and that she was the first person he told about his idea during the weekend. At one point in the speech, he describes wondering whether his idea had been tried already before: "If they had been used, I surely would have heard about it and so would everybody else including Jennifer, who was presently sunning herself by the pond taking no interest in the explosions that were rocking my brain" (Mullis 1993). Mullis therefore acknowledges that his girlfriend, Jennifer, was a competent, practicing scientist who would be able to let him know if he was going around in circles about something that had already been tried. At the same time, we are informed that Jennifer was more interested in "sunning herself," suggesting a particular kind of gendered self-absorption and also a focus on her body while he was making a brilliant discovery for humankind using his "brain."

At the lab on another day, one of the male community members was asked by one of the long-term (female) artist members if he would be coming to an event the next day. She offered him a free ticket. He said he was too excited about trying out his next experiment to attend her event. To this, she said something to the effect of "you have to be crazy to miss this opportunity which is only happening on this one day to do something that you can do at any time," making the argument that he was acting irrationally. That is, if someone "objectively" considered the options of attending something that only happened at a specific time versus doing something that could be done at their leisure, they would choose the first. Choosing the latter allows you to only do one thing; choosing the first allows you to do both in the long run. Love has often been said to be irrational, making you do things that do not make sense, and it appears that this scientist had fallen for science.

By the end of the twentieth century, scientists were perceived as less detached and more emotional without losing their credibility (Murphy 2012). How can the framework of passion coexist with the continued belief in scientists being objective and rational and the best arbiters of truth? Both the genius creator of PCR and the community lab scientist were swept away by the irrational love for science. These are also good examples of how the passionate scientist importantly remains a bit socially inept in certain ways—in particular when it comes to dealing with women (Turkle 2005; Willey and Subramaniam 2017). A degree of social ineptitude provides evidence of appropriate distancing from society, implicitly making these men more objective, better scientists.

This is not to say that men exhibited passion in DIY bio spaces and women did not. The story about Mullis and the creation of PCR was shared by a female instructor, who seemed to approach it with a "boys will be boys" attitude and a certain eye-rolling exasperation about the way men are socialized. At the same time, the story was retold as serendipitous and with a sense of awe and admiration for Mullis's genius even if it came at the expense of romantic relationship aptitude. Therefore, the proper gendered affect is not necessarily mapped directly onto members of any gender, and my observations suggested that many female-identified participants shared the appropriate attachment to science.

Yet women and people of color can participate structurally in maintaining exclusions that remain to be reinforced primarily along lines of gender and race. This is a key feature of neoliberal politics. The individualizing of experience and opportunities acts to allow a disavowal of the continued systemic, group-level divisions along which vast inequality thrives. Therefore, the larger question is not whether white women in these spaces challenged the defining ideas of DIY biology or not. What is important is whether the idea that certain groups of people belong in DIY biology is systematically challenged or reinforced, and further, whether inclusion/exclusion in DIY biology spaces means anything for larger inclusions/exclusions in the demos at large.

At first look, the gendered demographics of DIY biology spaces appears to challenge male-dominated sciences. The *Science* article that introduced the first community biology labs to the broader scientific community focused on two labs, both of which counted women among their founders. These women were prominently quoted in the article. Considering how inclusion of some members of a previously excluded group can act to disavow any continued structural aspects of exclusion (Hong 2015a), I was aware this likely did not represent the whole story. In fact, the late addition of some members of excluded groups can reinforce the idea that most of the members of those groups are fairly left out.

Despite the prominence of some women in public-facing media, the head of one lab lamented that a division of labor existed between women (besides her) working on art and the men working on science. A study of tinkerer demographics (Grushkin, Kuiken, and Millet 2013) found that 75 percent of respondents were male identified. There was no information about racial demographics because there was no question on the survey about race. My own interviews and observations along with media coverage indicates that white men continue to dominate these spaces and act as spokespeople for them, with some white and Asian women taking leadership roles in a few cases. The lack of interest in asking about race in the survey also points to a lack of interest in racial diversity, likely stemming from a white subject position that would be unlikely to consider or perhaps even notice the lack of diversity.

Why are DIY biology spaces dominated by white men? Are people of color and women less passionate about science or less interested in tinkering? And if so, what does that mean?

When I first met lab members across all sites, the question of gender equality came up quickly. When I described my project as feminist, interviewees often quickly translated this to mean that I would have questions about women's inclusion in the space. Instead of asking what feminist science studies is, lab members tended to assume that what I do is work to increase the number of women in STEM (science, technology, engineering, and mathematics). So, in my first meeting with founders of one community laboratory, they told me that they were not intentionally doing anything to increase diversity. Although one of them did express an interest in seeing more women in science as she lamented the gendered divide between artists in the space and scientists, at that time she did not think about doing anything proactive to change the demographics. About a year after those initial conversations, this same informant went for fundraising purposes to a STEM event aimed at increasing female representation in biology; after the event she returned to the lab excited about the high-profile corporations represented there and the possibilities to gain new funds.

I share this story (along with others here) not as evidence of an individual lab or person's motivations but instead to demonstrate where these lab formations fit into the larger socio-historico-political landscape. In the twenty-first century, signaling certain values has become profitable as corporations and governments negotiate maintaining a capitalist order where patterns of gender and race discrimination cannot be ignored. By "washing" one's image with superficial support of social justice ideas (e.g., corporate pinkwashing or greenwashing), an institution may avoid deeper concerns about the distribution of resources and power. For the lab founder I interviewed, feminism—or at least increasing women's representation— became a way to gain funding, thereby potentially increasing its priority level. The corporations sought to benefit from connecting their names to projects that demonstrated their own commitment to diversity and inclusion.

In another lab space, when I asked explicitly about their position and policies on racism and sexism, ethics, and inclusion, I learned about problems they had had with their first location. The group's members explained they did not initially have enough money to rent their own space, so they shared part of a computer hackerspace that was already operating. One problem that arose was that the computer hackers had pornography hanging up and made sexist jokes. One member explained it was not "family friendly." When asked to elaborate, he explained it was not the kind of place to which you could take your girlfriend, an explanation that seemed to acknowledge the space was unwelcoming to women. At the same time, the interviewee, who was a straight white male, was implicitly framing the space as a male-dominant one in which women would be present as girlfriends and therefore not as tinkerers themselves.

Although not exclusively the case, women did occasionally accompany their male partners to the space without participating. During one visit, I found out that a woman who was accompanying her male partner, a biohacker, to the space, was a corporate scientist and possessed extensive skills—more than her partner who was not a scientist by training. However, when asked about her participation, she explained that she did not engage in activities there but rather just came to hang out with her partner. Although drawing a simple conclusion for her lack of interest in participation would be misguided, I did wonder if she felt welcome in this lab considering the noticeable lack of women science participants in the spaces I visited.

In these conversations, a genderblind political context that still dominates much of Western society was evident. No one explicitly said they thought women should not be present, but there was also little emphasis on actively increasing gender diversity. Along with a stated belief in equality for women comes a belief that we have made great strides and have little work to do on that front. That is, my interactions and interviews in these laboratories demonstrated what has been called a postfeminist sensibility. Rosalind Gill (2008) reads what she calls a postfeminist sensibility in media culture as producing a simultaneous acceptance of a liberal feminism as commonplace and a "repudiation" of the need for a more radical feminist agenda. In one lab, members quickly noted the leadership of women in developing and running the space as evidence of gender equality. Members seemed to present this evidence as justification for their lack of interest in feminism. They seemed to use a strategy of distancing from computer hackerspaces that represented an insular geeky boy culture that was not "family friendly" to demonstrate the obviousness of a sexist-free culture in their labs.

A genderblind and colorblind orientation to justice ignores the historic inequities that have not been resolved. This is another place where considering the operation of gendered racial capitalism as a way of structuring society into a system of supposed natural human/Others (winners/losers, etc.) is useful. Including some members of a group into the global capitalist system does little to reverse any long historic divisions that were necessary to make this system thrive. Instead, for most interviewees, there was no need to have an explicit plan for increasing gender diversity.

However, one newer, young, queer-identified woman, Sam, shared how she was plotting how to bring another female friend with her so that she would feel more comfortable. Sam responded to our questions about diversity, racism, and sexism by talking about how she felt kind of awkward as a woman in the space due to a certain kind of "bro culture": "It's weird because biology is kind of stereotyped as more of a feminine science. So it's not necessarily that it's excluding other gender identities, but it's just that it's so default male. And, you know, I go in there, and they're friendly to me. They—I don't feel any discrimination against me, but it is default male, which is, I guess, a lot of scientific spaces, unfortunately."

Sam seemed to have a feminist orientation in that she felt that the space *should* be more welcoming and did not seem to make excuses for it or downplay the homogeneity of the space. She planned to deal with the discomfort by bringing her own friend, but she also was thinking strategically about how to address the problem in the long term:

> It's not something that I feel I have to address right now. And, since I am very new to this space, I don't think I have the clout to just say, "Let's just discuss this problem that, you know, specifically applies to me," you know, all of a sudden, you know. I feel like if I establish myself a little more then I'll have the power to be like, "Okay, these things—this, this atmosphere—this is why you have it this way, or something." But, like I said, there's nothing toxic. There's nothing aggressively against me, so it's not something that I have to address.

It was apparent to Sam that it was a (geeky) male space. That was not going to stop Sam from entering, but she was aware of it. She did not feel an immediate need to deal with it but did imagine addressing it in the future once she was more established.

Sam had spent time in various DIY spaces, not all focused on biology. She noticed that in the spaces with more women, the atmosphere was not necessarily better—not necessarily worse either, but different in the way that male members acted. In her experience, when more women were around the men felt more comfortable making jokes and comments that were sexist or homophobic. For example, she noted the use of "gay" for things they did not like. Her account of these spaces speaks to the postfeminist sensibility that may be heightened when a group feels that it has "dealt with" sexism already.

Another young woman new to one of the DIY biology lab spaces found it to be unwelcoming, although she did not identify any specific sexist, homophobic, or racist behavior. She expressed that the space was clearly for a "certain type of person" and referred to a sense of elitism that she felt there. Elitism was not one of the goals of community biology labs—quite the opposite, given their avowed focus on opening up science for all and the strong reference to amateurs being welcomed. How, then, could such differences have emerged?

Tinkerers Just Wanna Have Fun?

Having fun and playing with science is central to the tinkering culture of DIY biology labs. DIY biologists report a desire to explore the natural world in unfettered, fun ways (Kuznetsov et al. 2012). The terms *play*, *create*, and *explore* appear repeatedly on DIY labs' websites. Play is central to the dual (and related) objectives of creating a space for science to be fun so that more people can learn to love science and people who love science can have a place to freely explore.

Based on participant observation, Jane Calvert (2013) found that the idea of tinkering that denotes more playfulness and curiosity as opposed to more traditional engineering was popular in synthetic biology despite continuities with some engineering practices. Brigitte Gschmeidler and Alexandra Seiringer (2012) also found that the idea of playfulness was central to how the field of synthetic biology was framed in popular news reports (in German-language media). They suggested, "Playing was considered to be fun. Projects to produce bacteria smelling like bananas or peppermint (*Der Spiegel*, 14 August 2006; *Die Welt* online, 9 February 2009) conveyed fun as an essential part of the scientific work." Play was part of how the field defined itself.

Although I did experience bursts of excitement coming into the lab spaces and thinking about the possibilities of unfettered exploration, the courses did not exactly leave me feeling like these spaces were simply fun factories. One of the most important things I learned through the lab courses was how labor intensive the process of working with bacterial genomics was when you were DIYing most of it. Each of the classes was a mix of PowerPoint presentations and time working in the wetlab area. In the bootcamp course, we ended by sequencing our own mitochondrial DNA to look for ancestry markers. We isolated our DNA and sent it to another laboratory to receive the sequence, then compared the sequence with others using two different databases. In the other course, I learned the basic concepts of synthetic biology, which involved manipulating genes and splicing different combinations of original (or manipulated genes) together to produce organisms that performed new functions. The promise was that these new functions would be useful for society (and/or financially lucrative), but we simply practiced the technique by producing bacteria that would create a colorful pattern in the petri dish.

The labor-intensive nature of genetic DIY work was not lost on the community biologists nor on genetic scientists at large over the last fifty years. As one instructor put it, researchers in the 1970s and 1980s knew they needed new technologies if they wanted to go further in exploring genomes because "they realized they didn't have enough grad students, [clearing throat] slaves, to sequence the genome." This description of graduate students as "slaves" acknowledges the difficult and un(der) paid labor that the instructor and others provided through their apprenticeships in science labs. In the lab courses, although we did a lot of work by hand, I learned that one of the goals and premises at the intersection of DIY biology and synthetic biology was moving to greater automation. This goal seemed directly opposed to what I understood DIY work to be about: hands-on practices that unveil mysteries of how everyday things in our society work.

During my "Intro to Synthetic Biology" course, we watched a short YouTube video by Drew Endy, one of the self-appointed leaders of the field of synthetic biology.[12] In the video, Endy explains synthetic biology as "an approach to engineering biology" that builds on genetic engineering by adding three new foundational principles: (1) automated construction of DNA sequences, (2) standards, and (3) abstraction of complex biological systems. These principles mark a move toward more of an engineering model.

The DIY scientists I interviewed expressed the need for greater automation and credited greater automation for genetic science's successes such as the sequencing of the genome. One of my informants noted repeatedly that the really interesting stuff is the planning and then analysis—"not the middle boring part." They hoped to be able to get machines to do it. DIYers' access to biomedical experimentation has been perhaps ironically dependent on advances through Big Biotech that allow genetic sequencing to be completed more cheaply and quickly (Kelty 2010).

On the one hand, then, DIY has been associated with a move toward more local, hands-on understanding of the production and maintenance of the material worlds we inhabit—for example, making our own clothes, fixing our own cars, examining our own cervices, and growing our own food. It has been about gaining control over resources and knowledge so that DIYers are not beholden to big corporations. On the other hand, the DIY biology movement and its close association with synthetic biology would seem to be in ideological conflict: a DIY belief in opening the black boxes of how things work clashes with synthetic biology's drive to create more black boxes through abstraction and automation. The introduction of more engineering-based concepts has contributed significantly to this tension.

The field of synthetic biology represents an integration of engineering and biology, bringing together scientists who were traditionally siloed. One of the DIY synthetic biologists I interviewed used the metaphor of "barbarians at the gate" to describe engineers' relationship to the field of biology. This biologist, however, embraced the logic of engineer-based arguments regarding the need for standardization and automation of genetic research. The labeling of engineers as barbarians was a way to point out their simpleness and brute force approach to making things work. Another DIY synthetic biologist noted that most genes are not simple on/off switches and that really "frustrates the engineers." That is, engineers wanted to work with simple binary equations, but "nature" does not work that way. Yet each of these interviewees also explained how important it was for genetic science that engineers introduced the idea of standardization and automation. In this way these biologists occupied an interesting position where they disassociated with their traditional disciplines and were ready for the "new" direction but also saw the new innovators (engineers) as lacking an appreciation for the messiness of biological systems (nature). Apparently contradicting their own critique of engineers, they each explained the field of biology was too disorganized and needed standardization.

The engineering versus biology tension was not just about practical utility-driven exploration versus basic scientific exploration for curiosity's sake. It revealed gendered patterns of valuing certain kinds of scientific work. One of my interviewees said that genetic engineering of the past was more like "genetic arts and crafts" than engineering because everyone was off doing their own thing. The invocation of arts and crafts suggests a feminine (and less serious) pursuit, in contrast with engineering. Of course, the field of biology has also made the biggest strides in moving toward more equal gender demographics while engineering has continued to be extremely male dominated (National Center for Science and Engineering Statistics 2021).

Sam, a female-identified interviewee who wanted to work on fungi, was not sure what the practical application would be but said, "Eventually we want to observe how [plants or fungi] grow and how we can control how they grow." The idea of controlling biology for human interests was a common thread. She shared a common narrative about DIY biology in which the method of putting many heads together instead of one is better, not for an increased diversity of end goals but for getting the answers more quickly and finding things we might not without all that brain power. Even the artists, who were more often female identified, seemed to be interested in making biology work for profitable outcomes. There was a gendered progress narrative of doing more serious, profitable work instead of "genetic arts and crafts" of the past.

Though one of my informants focused on how the advent of automated technologies for genetic sequencing would help reduce reliance on graduate student "slaves," this large-scale automation also took the processing out of individual labs and into the larger processing laboratories of large companies. I found this out when we were working on analyzing our own mitochondrial DNA for ancestry markers. After we finished going through the basic steps of isolating our DNA, we closed the samples up in test tubes and sent them off to a company for sequencing. The instructor chose to FedEx it to a company that was just "over the river" in New Jersey so that if there were any problems we could go there in person. Perhaps this meant that, even though they were outsourcing the work, they did not want to send it overseas. The closeness also meant that we would receive the sequences in two to three days, which was important because our bootcamp course was scheduled each night for four nights in a row. We would receive the sequences in time to review them before our class ended. Indeed, we received the sequence data by email to the instructor before our final class session, and we were able to analyze them. The instructor noted the time stamp on when our genetic sequences were processed and wondered aloud about who was processing our sequences at midnight. If automation freed up graduate students' time, then who was running the machines and doing the work in the large corporate processing laboratories all night long?

Often, movement to further automation displaces more visible labor to other, less visible workers who work for less money (Atanasoski and Vora 2015). In the structure created by DIY biology, who are the ones doing the actual "hands-on" work? How was hands-on work being defined? What were the purposes of hands-on work?

From Genetic Hackerbatics to Business Incubation

For individual tinkerers, frustration set in over time when it was hard to make progress on projects, with some feeling that the lack of resources and dedicated time from other tinkerers held them back from creating profitable results to their experiments. Although those new to the spaces maintained an enthusiasm about finding these kinds of places, those who had been working for years exhibited real frustration about the lack of "real" scientific discovery.

One of the founders of a community biology lab believed that what was missing was advanced machinery. For this informant, this attitude seemed to emerge after years of work on promoting the idea of community synthetic biology hacker/maker spaces. The founder appeared exhausted, tired of what they said was merely "genetic hackerbatics" that were being done in the community labs. They described genetic hackerbatics as the time-intensive work of using enzymes to flip segments of DNA around; they believed it would be better to focus on coming up with the ideas then outsourcing the work of taking apart and putting DNA together.

Another founding member recounted initial meetups in people's apartments before they had a lab space, reminiscing and shaking their head at the same time about how they used their armpits to incubate DNA samples one night. (They noted that it was not a sustainable method.) I started to understand the significance of the word *entrepreneur* for many of even the most die-hard DIYers. They were ultimately interested in making "real" discoveries, which meant profitable, marketable products.

The tension between a rhetorical opposition to Big Bio and a desire to succeed materially in bio in big ways is particularly evident in the politics of property rights in DIY biology labs. Conversations about property rights have been central to the formation of tinkerer spaces. Drawing on computer hacker culture, DIY biologists have asked what open source means in tinkerer communities playing with biological materials. Genspace decided from the beginning that individuals would control property rights of what they created in the space. One "success" story for the space is the inventor of "OpenTron" (an automatic pipetting machine), who eventually moved to his own lab space to begin his own business. This entrepreneurial success puzzled me: How could an individual make money on something if it was open source? According to one of the course instructors, in some cases some parts of the plans are kept closed; in many cases, the business offers to create the product, but it is open in the sense that people can tinker with most parts of it. Therefore, these products have the possibility of being mass-produced. Open source had popularly earned a reputation of making software available for free, but those more familiar with the intricacies of open-source software report that many used this as an alternative form to make money. Open source was therefore not a wholly anticapitalist venture.

Tensions arose within these spaces due to the diversity of political orientations that members showed up with—many felt welcome by the idea of democratic science for many different reasons. Whether or not there should be any property rights given to individuals, what truly counted as open source, and when funded DIY biology projects stop being able to claim DIY biology status have all been debated in what some had referred to as "flame wars" because the conversations sometimes got heated and were taking place over the DIYbio mailing list. Two groups' decision to take sponsorship from Pfizer sparked considerable disagreement.[13] Some felt that as long as they maintained their own property rights, there was no ethical problem with partnering with this biotech company. Therefore, while making use of the popular anti–Big Bio positioning, at the end of the day these community labs (at least enough people with power to drive decision-making forward) embraced an end goal of creating real discoveries that ultimately would yield entrepreneurial success. Members of one lab told me the model they were striving for was that of the business incubator.

What Were the Limits of Unfettered Tinkering?

Tinkering and playfulness do not simply signal fun in the sense of "easy and care-free." These attributes also suggest that these labs are open for everyone to experiment in any way that they want. Even though DIY work may have been harder than one might imagine based on the image of the playful, fun tinkerer, there were other qualities that allowed the kind of entrepreneurial focus I found at community DIY labs. The lack of constraints was signaled on websites by the use of words such as "exploration," "creativity," and "innovation." This language gives the impression that anything you could imagine you could do in these labs. This is exactly what we were told when we asked about the rules on what kind of projects could be conducted at the labs. However, through further conversation we found a few examples of boundaries.

To try to understand the unwritten rules that must exist, I asked about how one could become a member and about the process for approving projects in the spaces. Members of one lab said that membership was decided through a very informal process and basically depended on who hung out in the space and wanted to join. I asked about whether there had ever been a time when they had to tell someone they were not welcome. They then told me about an incident when the unofficial head of the lab, who defined his role in their structure as a "benevolent dictator," punched someone whom he perceived to be racist and told him to not come back. When I asked what made the benevolent dictator decide that person was racist, I was told he kept talking about IQ and race and genetics. This understanding of racism was quite sophisticated in its acknowledgment of the power that doing certain research, asking certain questions, and producing certain scientific knowledge had in creating the conditions for race and therefore racism. Considering how often I was told that anyone could do any kind of project, I was surprised to find that there was a limit that included projects were considered overtly racist.

In the other lab, the hands-on lesson in our bootcamp class was to isolate our own DNA and run ancestry testing. This focus initially seemed to contradict the sense of the DIY biology community's antiracist attitude to genetic determinism and race. I shared my concern with two different instructors, asking whether this process would yield accurate information about ancestry.[14] I was assured that this was different than looking for race and that it was quite accurate. From this, I assumed that this group of biohackers had a different ethos about race and genetics, and later I was again surprised. While we stood cramped in the lab room taking turns isolating our DNA, one of the instructors shared their unequivocal disdain for the dangerous experiments being conducted in China's largest genomics institute, BGI. They knew about these experiments because they had been highlighted in the documentary *DNA Dreams* (van der Haak 2012). I later learned that the genetic research documented in the film was focused on finding genes specifically for intelligence.

The instructor's antipathy for BGI's research focus reflected the same concern I had previously encountered about unethical research on IQ, genetics, and race. Although the belief in the accuracy, usefulness, and unproblematic nature of ancestry testing seemed to conflict with these other concerns, I realized that there must be an important line between this research and the other. The slight move from race to ancestry makes ancestry tests more acceptable to scientists and the general public than trying to prove racial traits. The relationship between genetics and intelligence on the other hand must have developed a certain consensus that its racist history is inextricable from the question.

I was still troubled by the pushback I had received when I challenged the accuracy of these ancestry tests and the ethics of trying to determine race through them. The belief in DNA ancestry testing was so strong and so clearly related to perceived racial identification that during a session when we were comparing our ancestry results among us, the instructor continued to find new ways of comparing our DNA when my (presumably white European) DNA came up closer to the Asian woman in the room than the European woman. Eventually the instructor found a comparison and an explanation that fit their (our) racial assumptions. My observations matched others' findings that ancestry, as supposedly distinct from our racial classification systems, remains intertwined with race (e.g., Fullwiley 2007).

Conclusion

Cameron seemed to be correct in his assessment that there was not much happening in these DIY spaces that the FBI would be interested in. But the question remains: What are DIY labs producing? So far, the answer seems to be not much in a tangible sense.

There was no focus in DYI spaces on increasing demographic diversity, yet there was a political orientation toward appreciation and perhaps hope for more racial and gender diversity. Without explicit work in this area, it is unlikely that the situation would change. Without explicit racial or gender exclusion, the continued economic drive based in gendered racial capitalism means that those categories will continue to be operative. The culture of the labs' orientation to science and tinkering remained attached to typical Western science ideals despite the DIY, democratic rhetoric. These ideals continued attachment to different racial types (e.g., through belief in ancestry) and a rejection of feminized kinds of practices (e.g., dismissed as "genetic arts and crafts"). Although women were not wholly absent, those invested in DIY biology shared a similar attachment to masculinist, Western, and capitalist ideas of the goals of scientific practice. Sylvia Wynter's work has highlighted how changes in prominent scientific thought co-produced the racialized (and, I extrapolate, gendered) lines of inclusion/exclusion that define the human (Federici 2004; Lugones 2007). DIY community labs are creating continuities and discontinuities with previous versions of the human in a shift to a more passionate, democratic science practice.

INTERLUDE 2

If We Knew What We Were Doing

During my first year of graduate school, I attended our popular annual weekend retreat where graduate students and faculty gathered and shared beer, wine, trivia, and professionalization skills. At one point, we broke into groups to design a T-shirt, and then we voted on which design would become our official T-shirt for the year. My group won. Our final T-shirt design included the phrase "If we knew what we were doing, we wouldn't call it research" along with a drawing of a synapse with neurotransmitters represented as floating bubbles with pictures of our faces inside them.

At the time, I remember not understanding what the phrase meant. I had a feeling I should have known this, based on how quickly the others embraced it and laughed about it. There seemed to be some kind of inside joke, and everyone else seemed in on it. So I went along, but I kept wondering. What was funny about the phrase? Was there a play on words I was missing? Was it funny because people think we know what we are doing but really we are just messing around and trying things? It did not seem funny to me, but as a female-identified graduate student at the time I was trying to fit in. I wanted to be considered smart and to get it.

Recently, as I was reflecting on that retreat, including the fun and playfulness of our T-shirt design and the overall energy of the program, I wondered again what the phrase meant, this time from my new position in feminist science studies. I went to the internet ask Google about the meaning behind it and who may have said it. I quickly found that it was attributed to Albert Einstein, but I also found that there was no original citation for the quote. That is, even though it is used as an epigraph in science textbooks and science journal articles (and printed on T-shirts with Einstein's picture), you cannot find an actual source where Einstein wrote it or said it.

This, as a side note, made me feel better for the confusion I felt that day as I hoped no one would figure out that I had no idea what the phrase meant. Now I am sitting back laughing at myself for my anxiety in that moment over a potentially misattributed and at the least unverifiable quote. However, what I find most interesting now is that this quote fits the modern scientist figure of "the modest witness" and the newer figure I will examine here, "the tinkerer."

There is modesty in saying that we do not know what we are doing but are searching for answers. Tinkering also assumes that we cannot predict the result and instead are simply messing with things and trying stuff out. Both approaches to knowledge production, importantly, remove us and our positionality from the equation of observed results. I explore both of these constructs in the next chapter.

CHAPTER 2

The Tinkerer as a New Scientific Subject

Community biology labs have had little success in producing actual new scientific products or discoveries. However, the spaces were important for what and who they represented and what they helped to produce: a new scientific subject, *the tinkerer*.

DIY biology spaces arose from the context of both existing DIY movements and histories of scientific discovery. In these physical laboratory spaces, recent interest in makerspaces and hackerspaces has intersected with biological sciences. Hackerspaces were first exclusively the domain of computer geeks while the focus of makerspaces has spanned from electronics to textiles. However, each term, hacking and making, has found more widespread use with intersecting meanings in the last few years (Tocchetti 2012). Although some have argued that hacking was explicitly an anticapitalist, antiauthoritarian project based in a passionate approach to work (Himanen 2001), Graeme Kirkpatrick (2002) argued early on that there were significant intersections between the protestant work ethic and a hacker work ethic, thereby situating hackers within American values of capitalism and not necessarily an oppositional force. DIY biology emerged into and from this complicated, contested history.

Hackerspaces and makerspaces have been criticized for their overall lack of diversity (Masters 2018; Toupin 2014), and biohacking specifically has been critiqued for the ways that its antiauthoritarian ethics do not preclude entrepreneurial goals (Delfanti 2011). These spaces' racial and gendered effects in particular intersect with capitalist goals—a relationship that DIY biology spaces are uniquely positioned to illuminate.

From Modest Witness to Tinkerer

Feminists and other critical scholars have long critiqued notions of pure objectivity based on the observation that Science itself could not exist without Scientists. Within broader cultural shifts we see a co-production of different orientations to science and new requirements for who counts as a proper scientist. Many have

TABLE 2.1

HISTORIC SHIFTS IN SCIENTIFIC OBJECTIVITY

	TIME PERIOD (CENTURY)			
	17th—mid-19th	*Mid-19th—mid-20th*	*Mid-20th—late 20th*	*Late 20th—present*
Epistemic value	True to nature	Mechanical objectivity	Trained judgment	Objectivity contested
Scientist figure	Modest Witness		Expert	The Tinkerer
Attribute	Trustworthiness	Detachment	Intelligence	Passion

Note: Rows 1 and 2 are excerpted and modified from Bero (2019).

noted there have been varied and even conflicting definitions of objectivity as an epistemic value over the course of Western history (Roy 2008).

For at least 200 years (from the seventeenth to mid-nineteenth centuries), the "modest witness" was the dominant figure of the modern scientist (Table 2.1). The modest witness had to observe the natural world from an objective standpoint, simply bearing witness to it. This supposedly neutral standpoint was coded as white, male, and moneyed because only certain people could be trusted to view nature in this way (Haraway 1997). This scientist in practice was neither modest nor merely witnessing "nature." According to Lorraine Daston and Peter Galison (2007), the scientific self who was produced at this time was that of the creative genius who was needed to properly abstract truth from his observations of the natural world. Not surprisingly, gentlemen were considered to be the most trustworthy for this task.

In the mid-nineteenth century, the biological sciences became more important in assessing what was true about our natural world and who should rightfully make decisions about it (Wynter 2003). Challenges to the gentleman and his supposed objectivity were answered in part by recourse to new technologies such as photography, leading to a move away from trust in the elite and toward the objectivity of machines. Daston and Galison (2007) identify this period as marked by the epistemic value of "mechanical objectivity" (Table 2.1). In this time period, the proper scientist was supposed to be detached from the results, and technology was heralded as a way to ensure proper detachment and objectivity. However, technologies/machines do not operate on their own, so subjectivity continued to be a problem.

This period included the now widely refuted racial sciences, including eugenics. Scientists at the time looked for racial differences by relying on the science of measurement. As Evelynn Hammonds has noted, "If we just take African Americans as an example, there's not a single body part that hasn't been subjected to this kind of analysis" (Adelman 2003). Along with Hammonds, Stephen Jay Gould

(1996), Londa Schiebinger (2004), Anne Fausto-Sterling (2000), and many others have provided extensive histories showing the "mismeasure of man" is not simply about an overreliance on machines causing errors but dominant ideas on gender, race, class, ability, and sexuality circularly reinforcing ideas of the inherent hierarchy of white men.[1] Included in these supposedly neutral analyses of human difference was the concept of innate intelligence, which became most popularly associated with IQ tests.

The idea of natural intelligence is closely related to that of rationality, which has been crucial to coloniality. The construction of the reason/emotion binary, aligned with oppositions such as male/female and white/non-white, began with fifteenth century Western colonization (Anzaldúa [1987] 2012). This does not mean that men's actions have been absent of emotion, but rather that white men's emotions are not coded as irrational but instead normalized and often invisible. This ability to reason that was the domain of white, property-owning men was coded through the concept of intelligence in the nineteenth and twentieth centuries (Gould 1996).

Natural intelligence became an important measure of who could become a scientist—as well as structuring much of society. Around the seventeenth/eighteenth centuries, the university system split into schools of humanities and sciences, thereby reinforcing the colonial distinction between reason/emotion and corollary binaries such as rational/irrational and culture/nature. The Sciences became valued as the best way to discern truth (Wallerstein 2004). By the middle of the nineteenth century, biological sciences had moved to the top of the epistemic ladder, meaning that who controlled biological knowledge was important for all kinds of societal decision-making and a social ordering where capitalism made "natural" sense. Through the circular logic of defining intelligence through racialized and gendered metrics, and then using intelligence to make supposedly unbiased assertions about who was fit to be a scientist, the scientist cohort remained mostly homogenous (white, male). Perhaps in part due to the major critiques of racist sciences of eugenics, by the mid-twentieth century there was shift to "trained judgement" (Table 2.1) in which properly credentialed scientists were responsible for interpreting data (Daston and Galison 2007).

In the mid-twentieth century, changes and challenges to knowledge, power, and politics, allowed a more passionate figure of the scientist to emerge (Murphy 2012). Pure objectivity has now been roundly critiqued for not only whether it is being met but whether it is possible at all. Feminists among others have asked whether this ideal, even if possible, would be desirable, offering instead other epistemological stances such as standpoint theories and situated knowledges. This shift marks a historic moment where objectivity is highly contested. This is true across the political spectrum in the United States. Patient-centered movements, such as the feminist women's health movement and AIDS movements, have partly paved the way for a new look at science and scientists. Another dominant understanding in this moment is that historically the concepts of IQ and natural intelligence have been used to harm people of color and women. In chapter 1 we saw this belief reflected through the comments and actions of DIY

THE TINKERER AS A NEW SCIENTIFIC SUBJECT   57

community biologists. More broadly, pushes to remove standardized tests such as the SATs are gaining momentum due to the understanding that these "standardized" and supposedly neutral tests are quite biased.

Relatedly, in the last couple of decades the supposed division between basic science and technologies or applications has become more difficult to maintain, leading to the use of the term "technoscience" by critical scholars to encompass the fields previously thought of as basic and applied sciences/technologies. This is the historic moment in which DIY biology emerged. DIY biology blurs the line between biology and technology by bringing the fields of engineering and biology together. It also challenges academic sciences by claiming that anyone with a passion for science should have a right to it.

For the tinkerer, as distinct from previous figures of the scientist and their relationship to objectivity (Table 2.1), knowledge production is considered at its best when the scientist takes an active, creative role in it. The tinkerer locates value in "playing with nature," apparently flipping the binary of reason/emotion. As chapter 1 details, this has not automatically resulted in a more racially and gender diverse body of scientists in the labs. How, then, do these spaces remain exclusionary? If in earlier periods exclusion was based on the binary notion that some subjects were too emotional to reason, what is the exclusion based on today? What are the continuities and shifts from the modest witness to the tinkerer? Who is best positioned to be able to play with nature?

The tinkerer has become most prominent as a figure at the intersection of synthetic biology and DIY biology. The tinkerer has previously been proposed as a model in more applied areas of science and for engineering (Caporael, Panichkul, and Harris 1993), but today's emergence of the tinkerer as a biological scientist signals a new kind of understanding of basic science research. There have been many challenges to Western sciences, such as criticism of their elitism and exclusions in who is a proper scientist as well as what kinds of questions are asked and why. Across the political spectrum, critics have questioned whether basic sciences are useful for our everyday lives and world. The tinkerer emerges in part as an answer to those concerns while maintaining the supposed objectivity or special access to truths of previous scientists.

The community biology lab BUGSS took pains to announce on their website that their "space [was one] where it is ok not to know." This orientation can signal that one can come as one is, without previous experience or knowledge. It also suggests a way of doing that starts without preconceived notions. This matches with common definitions of the tinkerer who tries things out without a clear goal. In this case, the tinkerer messes with biological material as the method to discovery and innovation. Embedded in the approach of experimentation without a goal or preconceived idea is a claim to being unbiased. This shift, however, is dangerous. It appears to answer challenges that science is inattentive to the public's needs while continuing a similar kind of exclusionary science and innovation.

The rationale for why certain people are better suited to be observers has changed over time, along with which epistemic values are most important (Daston and

Galison 2007). The moves from one belief to another over time did not mean there were sharp changes in which one belief system was wholly rejected and replaced by another. Often the idea that rose in dominance served to answer previous critiques, appearing as a progressive shift to gain more accurate and more objective truths. Although more objective forms of science that supposedly did not rely on the economic status of the trusted observer should have displaced the modest witness, this is not what happened in practice. What we see in these small shifts in who is trusted as a knowledge producer is that white, wealthy men continued to be overrepresented as scientists and the knowledge produced continued to reinforce systemic social inequities, benefiting the owning class. This phenomenon speaks to the flexibility of gendered racial capitalism.

Gendered racial capitalism works by maintaining group differences that mark groups as more or less deserving of economic control. This trick of counting certain groups and therefore certain people out of economic opportunities can only last so long within its own capitalist logic of economic freedom for all. Therefore, there must be flexibility in the system that reacts and regroups based on challenges to exclusions.

Biology developed as a rational way to divide up those deserving and undeserving of power, deciding not only whose lives literally matter (matters of life and death) but also, importantly, who belongs within the ranks of scientists. Those who count as scientists, in a structure where science holds such epistemic power, have a lot of power to maintain or challenge the logics of group difference. Therefore, maintaining control over both the purposes of science and those included in science is important for maintaining gendered racial capitalism. Most recently, apparently more inclusive definitions of science and scientists as embodied in the figure of the tinkerer have produced similar patterns of exclusion and have maintained adherence to the idea of individual economic success as the ultimate goal of inclusion.

There are three interrelated qualities of *the tinkerer* in this historical moment: (1) a naturalized discourse of what makes the tinkerer a scientist through passionate affect for science, (2) a discourse of disenfranchisement from scientific knowledge production, and (3) an assumed end goal of capitalist gain reached through entrepreneurial spirit.[2]

FROM INTELLIGENCE TO PASSION FOR SCIENCE

One of the founders of the DIY biology movement, Ellen Jorgensen, is cited in the *Science* article as saying that "scientists are born tinkerers" (Kean 2011). Another important founding member of this DIY biology movement, Eri Gentry, argues that institutional lab spaces stifle innovation:

> Although not a biologist herself, BioCurious co-founder Eri Gentry says she hunted down lab space to rent because biology students she knew through BioCurious had grown weary of pursuing narrow Ph.D. research topics and wanted

THE TINKERER AS A NEW SCIENTIFIC SUBJECT   59

to tackle side projects they were passionate about. The setup in most science labs today "doesn't breed creativity," she argues.

That's a common sentiment in DIY bio, and it motivates much of the passion. Scientists are born tinkerers, says Jorgensen, also an assistant professor of pathology at New York Medical College. "This place [Genspace, a DIY bio community lab] is made for spare-time tinkering." (Kean 2011)

The phrase "born tinkerers" connotes an innate (perhaps genetic) difference that marks one as a tinkerer, and the idea of "breed[ing] creativity" points to an environmental impact on tinkerer development. Such a complex gene–environment interaction relies on the idea of naturalized types of difference. The concern that traditional labs do not breed creativity suggests that tinkerers need more supportive spaces to rightfully innovate and explore.

The idea of types of genetic traits that make people more likely to do something under the right conditions (environment) represents a common way of understanding inequities in occupations and educational opportunities without acknowledging the societal structures that produce such disparities. For example, entrepreneurialism, like much capitalist success, has enjoyed the myth of both meritocracy and innate biological disposition combining to produce success. In the case of entrepreneurialism, a theory prevails that successful entrepreneurs can thank a genetic predisposition to "take risks." However, evidence shows that the marker that predicts success is class status, and those who are most likely to succeed come from wealth; their families' ability to support them and provide connections to other wealthy investors is the precursor allowing some people to take risks in business (Levine and Rubinstein 2013; Xu and Ruef 2004).

Gender and race are coded into the notion of the tinkerer as a biological "type," along with interrelated notions of disease and personality. Although emotionality and subjective attachment would seem to be in opposition to producing objective, detached scientific knowledge, the passionate, entrepreneurial, driven scientist emerged in a post-1960s reconfiguration of the moral and social politics of knowledge (Murphy 2012). Angie Willey and Banu Subramaniam (2017) trace this figure further to argue that the scientist, traditionally celebrated as socially detached, is brought into the sphere of the "social" through a process of naturalization. This new geeky scientist represents a shifting white masculine subject who is passionate about science, thereby making him more "social"/"normal" at the same time as being able to remain socially unaccountable through the naturalization of an "asocial"/"genius" brain. The subject of community bio labs fits this merging of passion and entrepreneurialism. The idea of an emotional attachment to science—that might even foster irrational decisions to prioritize one's love of the discipline over other social values—constructs the newly naturalized proper scientist, the tinkerer.

The *ability to do science* is often thought to be based on an innate intelligence. This belief is reflected in ideas of being "smart enough" to do science and the continued use of standardized tests to gain entry into the nation's best science schools. Decades of critical scholarship have demonstrated how intelligence is racially coded and how that has had an effect of legitimating the exclusion of racial minorities and women from the sciences. Due to this research, even though the idea of a natural aptitude for science remains,[3] it has become less popular. This point was forcefully illustrated in one of the community biology labs I visited when a man was punched for talking about race and intelligence. There was no tolerance for researching or discussing ideas that reeked of eugenic discourses—at least around race and IQ.

The discourse of *passion to do science* is ostensibly a gender- and race-neutral measure for inclusion in science. This reasoning appears in science literacy and science diversity projects alike (e.g., "Physics Is Phun" classes). This is the discourse mobilized in the production of the subject of the tinkerer. This discourse appears to open the possibility that anyone can do science at the same time as interest in science becomes a prerequisite. That seems reasonable enough—if you are not interested in it, you should not do it. But the epistemic power of science has the potential to foreclose other knowledge-producing systems as legitimate and to exclude those who do not feel comfortable in the space as possibly not having that natural desire/passion. The use of the correct affective relationship to STEM fields (science, technology, engineering, and mathematics) remains a way to universalize and create a colorblind, genderblind, nonpolitical way to avoid questions of structural inequalities (Slaton, Cech, and Riley 2019).

Defining the tinkerer through a seemingly innocuous desire for science may appear to fit into the idea that community labs open up access to science for "everyone." However, *desire* to do science (or to tinker, in this case) as a prerequisite can act to exclude. The notion of a desire for science may reinforce a gendered division of who really belongs based on who cares more deeply about science.

The idea of natural aptitude, then, has not been totally displaced by a discourse of passion for science; instead, there has been a conflation of the two. Affective desire has not totally replaced the idea of being "smart enough" to do science and math, but it has become a seemingly fairer way to delimit the boundaries of scientist. In science education a similar effect has been documented in the shift away from IQ toward measuring "interest" in science (Kirchgasler 2019). Kathryn Kirchgasler (2019) has compellingly shown, however, that the assumption of different groups having differing natural "interests" has been around for at least a century, where this acted to determine different education, career, and life paths. Therefore, the shift to interest (or passion) for science should not be uncritically embraced as progressive. Passion (or interest) for science continues to naturalize the idea of the proper scientist, continuing to create categories of those who can and those who cannot—those who are meant to do science and those who are not. This naturalization of desire for science creates a rationale outside social injustice for why some people are not included as knowledge producers and maintains Science's ultimate epistemic authority as the best way to produce knowledge about our bodies and environments.

In the case of the tinkerers, there is an implicit claim that at the same time as science is for "everyone," those who actually belong—that is, those who demonstrate an affective desire for science—become naturally separated from those who

THE TINKERER AS A NEW SCIENTIFIC SUBJECT 61

do not have the passion for science based on supposedly race-blind and gender-blind terms.

DISENFRANCHISED FROM TRADITIONAL SCIENCE

The second defining characteristic of the tinkerer is disenfranchisement. This is exemplified in the title of the *Science* article "A Lab of Their Own" (Kean 2011). The idea that tinkerers need space of "their" own places them within discourses of unjust exclusion.

How does a group of mostly white, male, and relatively well-off people make a claim of oppression? Co-optation of exclusion discourse for white people in the United States is not uncommon. Rooting for the underdog and seeing oneself as overcoming adversity through hard work is part of a common American narrative. This idea is captured in a 2010 *New York Times* article about George Church, whose photo made a significant appearance in the Glowing Plant project's fund-raising video (Duncan 2010).

Church is one of the most public geneticist personalities and a leader in the field of synthetic biology; in the article he narrates how he came to genetics through a tale that combines innate passion for tinkering with the story of an underprivileged American male who worked hard to overcome disadvantage. Church says, "I always loved computers—it's something inside you." This self-description reinforces a common story that positions computer tinkering as a precursor or closest analogy to the synthetic biology movement. Part of this tale is that anyone (who is a white male with a garage) can innovate and become the next Steve Jobs.[4]

The love for computers (things) as a different orientation than love for women (flesh) has been prominent in male-dominated computer hacking culture for decades (Turkle 2005). Church not only presents his "love" for computers as something inside of *him* but makes a claim that this kind of passion is more generally locatable in bodies, thereby naturalizing once again the passion to tinker. In this article, however, he goes further and says that while he had this innate passion for computers, he did not have access to computers, so he had to make one himself—calling upon the powerful American mythology that this country is great because of underdog, hardworking white males.

The next part of the story, however, is that his mother married a physician, and he then became interested in biology. Considering the unlikeliness of cross-class heterosexual relationships,[5] I was surprised that this hardworking boy who had to build his own computer was of the physician class. The disjuncture between the class reality of white people and their perceived disenfranchisement is quite common (Reeves 2017). I looked further into his story and found through other interviews that his mother, Virginia Anne Strong, was an author, attorney, psychologist, and architect who seems to have divorced and remarried twice; her final relationship was with Dr. Gaylord Church, who became George's third dad (Temple 2014). A mother trained as an attorney does not appear to be part of a typical working-class story either.

George learned what he could about computers—not exactly "against all odds"—at Academy, a preparatory boarding high school in Massachusetts (Nair 2012), which today costs more than $60,000 per year. George Church's access to wealth does not mean that he could not help to democratize science, but his story's apparent contradictions regarding material and cultural access to privilege speak to the prominence of the claim of disenfranchisement. This is not to say that everyone in DIY biology or synthetic biology comes from an upper-class background; rather, Church's personal story exemplifies the narrative of the biohacker who works against odds to succeed due to passion and perseverance.

In the case of biohackers, as Christopher Kelty (2010) has pointed out, this group carefully positions itself as "outsiders" at the same time as being very much part of the "included" or "inside." Tinkerers freely admit to the fact that most of their founders and about half of their members are or have been part of traditional biosciences or engineering programs (Grushkin, Kuiken, and Millet 2013). For the tinkerers I interviewed and spent time with, there were no qualms about being insiders or even receiving funding from big biotechnology. This connection jarred against the language on their websites and in news articles, which seemed to frame the tinkerers as countercultural subjects.

Thus, the story of lack of access (outsiderness) and the reality of insider belonging seem to coexist without causing concern for the tinkerers' claim to need "labs of their own." The reason biohackers give for needing this "outside" space is that there is a lack of freedom to explore projects that they might find interesting because of traditional funding structures (government grants and corporate money). Central to this material relationship is the need to innovate.

As with the raised fist symbol for DIYbio, the kind of revolutionary tradition this disenfranchised claim draws on is that of social justice movements. An image of the Boston Tea Party or the Gadsden flag ("Don't Tread on Me") might instead be used if a direct connection to the American Revolution was the goal. Instead, this choice of a revolutionary fist calls up racial and gender justice movements, and particularly those rooted in anticapitalism because the fist was first used in socialist movements: both the Black Power and feminist movements of the 1970s were deeply concerned about fundamental problems with capitalism.

The use of the fist along with the larger claim to need space to freely do science represents an appropriation of discourses of disenfranchisement and the need for freedom without being rooted in deeper social justice ideas. The goals of this revolution are not for the eradication of capitalism—perhaps quite the opposite.

Entrepreneurial Capitalism as the Natural End Goal of Democratic Inclusion (or the Flexibility of the Tinkerer and Capitalism)

In DIY community labs' self-descriptions there tends to be an equally strong emphasis on science for fun and science for entrepreneurial purposes. Fun and capitalism find a relationship through entrepreneurialism. This pairing works to

THE TINKERER AS A NEW SCIENTIFIC SUBJECT 63

further naturalize capitalism: pairing fun with science produces the tinkerer with a natural drive to explore and discover. The capitalist spirit of innovation for profit is naturalized so that it becomes *the* end goal of democratic inclusion in Science.

Innovation has become synonymous with Science and human progress in recent years despite strong evidence that this may lead to neglect of other human activities such as maintenance and care (Vinsel and Russell 2020; Wisnioski, Hintz, and Kleine 2019). Combined with this neglect, capitalist-driven innovation has been criticized for how directly destructive it has been, such as in its environmental impact around the world. For example, the claim "A Nursery for Explorers and Entrepreneurs" appeared on the website of Genspace, one of the community synthetic biology labs, making an association with childhood through the use of "nursery" to naturalize and make innocent their projects of exploration and entrepreneurialism.[6] In reality, each of these relates to quite politically non-neutral practices. Specifically, exploration is most associated with colonization, in which the presumed active subject is centered in discovering geographic regions, peoples, and/or practices that already exist. This centering means that the explorer's point of view is what matters for decision-making. Entrepreneurialism, of course, is a form of capitalism. Neither are politically neutral practices, and both can be associated with exploitation. However, through the proximity to childhood and the idea of nurturing these types of people, explorer and entrepreneur become naturalized as innocent positions. The notion of innocence also implies that this group needs to be protected and properly nurtured, which is part of the implicit justification for why tinkerers need a "lab of their own."

By the middle of the nineteenth century innocence had become linked to childhood (Bernstein 2011). Robin Bernstein argues that the innocent child is a racialized figure—a white child. She argues that this figure had the power to transfer its innocence "to surrounding people and things, and that property made it politically usable." The white child was used together with the image of a "happy slave" to make slavery politically innocent. This happens through a process of naturalizing the idea of childhood innocence and then transferring that naturalization to the political issue—for example, slavery. Bernstein further argues that the white child has a quality of "obliviousness" that is part of what creates the innocence. Similarly, through the discourse of fun and also at times a more direct reference to childhood and youth, the tinkerer figure is imbued with the innocence of the white child. This innocence and obliviousness is manifested in an adherence to scientific objectivity at the same time as the tinkerer tinkers for economic gain, thereby further transferring the innocence to the entrepreneurial project.

Passion for tinkering is further imagined to be removed from "work." Tinkers, whether or not they receive their paycheck from Big Bio or government grants to their university labs, do their passionate work, their "spare time tinkering,"[7] outside these spaces as a hobby. This obscures the role of the economic objectives while tinkerers' entrepreneurial spirits drive the science being done as hobby. The end goal of "fun" tinkering, however, is the ability to "innovate."

The end goal of the tinkerer rebellion, then, is the freedom to become entrepreneurs. Since the nineteenth century, the prevailing definition for the human has been *Homo economicus* or "bio-economic Man" (Wynter 2003), in whom modern concepts of rationality, liberalism, and biological determinism merge. That is, the purpose of being human and the thing that makes one human is circularly tied to one's ability to be an economic subject based on whether one is capable of making rational decisions. This makes it unnatural and impossible to consider any reason for acting outside of an economic drive (Brown 2015). This is also a necessary condition for being considered a political subject. Without economic interest, a claim for inclusion in the demos becomes unintelligible in our current world system. The engineering/biology tension highlights how discovery for fun and for profit can coexist, producing the seemingly contradictory antiauthoritarian, procapitalist tinkerer: the more "basic" biological orientation to exploring nature for enjoyment and discovery for fun (and playfulness) meets the practical engineering orientation of solving problems for profit.

Lilly Irani (2015) found a similar result through ethnographic and historical work with computer designers who participate in hackathons in India. She argued that while the goal of hackathons is to produce new technologies that can help change the world for the better, they do not often produce a viable product. However, they always do produce a certain entrepreneurial citizen who mixes desire for social change, nationalist politics, and a complicated relationship with the idea of democracy. In the postindependence context Irani writes about, there is a celebration of the individual entrepreneur who cares about saving the country but is also wary of the slowness of true democratic processes. Hackathons are celebrated as participatory but are limited to certain kinds of people with technical backgrounds and time to innovate.

In many ways the tinkerer is not a departure from past figures of the scientist, such as the modest witness. The privilege of the modest witness is ignored or rather obscured as he becomes a literal role model in some cases. Tinkerer and maker movements at large have resonances in earlier maker movements based in an American exceptionalism of white men's (literal) hands—the combination of their work ethic and creativity against all odds (Foege 2013). These projects often call on "a long [American] tradition of self-reliance and small-scale technology, attainable by 'everyman'" that traces back to "the 'republican gentleman' and 'independent producer' ideals prevalent in the early nineteenth century" (Dunbar-Hester 2014).

Within DIY science, Charles Darwin is praised as an example of how anyone could follow their interest and do science back in the day (Biochemical Society 2017). Darwin, in fact, was rich and funded his own explorations. He was not an employed scientist but rather a gentleman of means who could be said to have "tinkered" with/in science. Darwin, while experimenting at a moment where a value on laboratory work and mechanical objectivity were beginning to replace the modest witnesses of earlier centuries, carved out a strong place for himself as a wealthy man working in his home (White 2009).

The idea that anyone could do science in the good old days harkens back to a structure that made the scientific field quite elite due to the money necessary to participate. Although there is more economic diversity in science fields today, this remains an issue due to the high costs of education as well as cultural forms of exclusion based on race and gender.

THE POLITICS OF DEMOCRATIC INCLUSION (AND EXCLUSION)

What is at stake in the co-optation by tinkerers of the politics of inclusion of a disadvantaged group is (1) a further racialized and gendered exclusion from the means of scientific knowledge-making through a colorblind and genderblind politics and (2) a definition of democratic science that leaves intact participation in capitalism as the natural and most desirable goal of inclusion. The tinkerer, then, is a subject formation of gendered racial capitalism. A seeming democratization of sciences in this political moment offers a shift in the definition of a proper scientist at the same time as reconsolidating a hierarchical position of those who are the proper subjects of scientific knowledge-making for the purposes of entrepreneurial gain.

Inclusion in the demos for the last few hundred years has been limited to those with appropriate economic status—*Homo economicus*. Race and gender have been central to how those deemed worthy of inclusion as "economic man" have been determined by creating categories of difference that are used as justification for exclusion. The belief in biological determination of types of people and the ability to pass on traits yielded what Sylvia Wynter (2003) calls "bio-economic man." Bourgeois (European/white, middle–upper class) women have been expected to do unpaid reproductive/domestic labor while people of color and poor people across genders have been expected to do free (slavery) or low-paid labor for the benefit of capitalists (Federici 2004; Wynter 2003). In the context of DIY labs, nonserious work (e.g., "genetic arts and crafts" or art-based DIY biology projects) and more rote, noncreative labor (e.g., genetic sequencing late at night) are gendered and racialized ideas, regardless of the exact identities of those filling the roles, and at the same time are more likely to be filled based on long-standing societal categories.

A passion for science as a prerequisite for inclusion is skewed toward white, male heterosexuality. This is not altogether surprising because tinkerer identity is closely linked to a broader geek or nerd identity. Geek identities are often presented as if they are universal, colorblind, and genderblind categories based on intelligence and interest in certain topics. However, these requirements themselves are defined through white masculinity, thereby rendering particular bodies and actions unintelligible within this frame (Dunbar-Hester 2014; Eglash 2002). A lack of diversity in DIY community labs and the participants' political beliefs around gender and race diversity help to illustrate how these categories of difference continue to be used to maintain an inequitable distribution of resources as well as life chance (gendered racial capitalism). Clare Jen (2015) points out that the figure of DIY biologists comprises those with masculine roots such as the garage biologist and asks why DIY biology figures such as "Ms. Science" and "the feminist biohealth hacker" are not as prominent.

A particular public is formed here despite claims of inclusion for everyone. That public is one that loves science, has a drive to explore and tinker, and thereby has the (capitalist) spirit to create and profit. "Passion for science" operates as a new way to legitimize, through recourse to naturalization, unequal access to scientific knowledge production. Passion and excitement for science are connected to the idea of "playing" and having "fun," which further attaches to the idea of innocence, childhood and thereby to detached, objective scientists and science practice. Although fun, playful experimentation may seem like a challenge to an economically driven focus on producing marketable/translatable science (e.g., pharmaceuticals), the way fun circulates does not break the rules set up by the dominance of *Homo economicus* (man who is defined primarily by an economic drive). Fun, playful exploration is not in conflict with entrepreneurialism but instead is seen as an almost necessary precursor. The "freedom" to explore is thought to produce more creative final products and answers. Success is measured by the marketable end products of discovery.

These ideas are not unique, of course, to synthetic biology or biohacking community labs; they can be seen in the broader cultural shift in new businesses and entrepreneurism that allows business-men-in-the-waiting to work more freely in their T-shirts and jeans instead of stuffy suits. The "success" of this model has been exemplified by Google: workers are paid handsomely at the same time as being given flexible work schedules, gourmet food, and games available at the office. This has been part of the technology revolution. Now young white men do not have to go off campus to play golf but can play pool at the office![8]

In genetic labs (community and otherwise) the work is outsourced. These outsourced jobs created through synthetic biology's further reliance on "automation" are presumably underpaid and become a form of feminized and racialized invisibilized labor. As Neda Atanasoski and Kalindi Vora (2015) point out, the belief that technology has moved us to a posthuman network obscures a new labor politics of invisibilized technological labor. The invisibilized labor of supposed automated genetic processes is left to Othered workers.[9]

A transition appears to have taken place over a relatively short period of time from an ethos that was in part about a hands-on approach to science and the idea that anyone could engage with few resources (Roosth 2017), to one in which labs vied for grants, wished for more expensive equipment and more dedicated participants, lamented the lack of "real" discovery happening, and even considered and sometimes accepted sponsorship from biotech corporations. This transition is perhaps better understood as a tension that existed from the beginning and continues to heighten. The tension is a seeming contradiction in foundational ethical values.

The idea that we can understand biology differently and perhaps better depending on our relationship to the matter itself has been a popular point in feminist biology and more mainstream biology over the last few decades. Automation appears to move us away from that kind of relationship and the epistemological and methodological benefits claimed through it. In a critique of the direction that biological science labs are moving in by outsourcing work, David Smith (2014) cites Barbara McClintock; her biography, *A Feeling for the Organism* (Keller 1983), offers examples of insights that likely would have been lost had the scientist been outsourcing the labor. Smith also fears that the push for commercialization of science is causing the end of academic goals as we know them—the pursuit of pure discovery through creativity and hands-on experimentation.

DIY biology at first glance appeared to be responding to the critique that Big Bio is killing creativity. However, the focus on freedom from Big Bio and the academy did not mean that community biologists wanted to actually get closer to biological matter. In fact, the common thread was about controlling biology for human benefit. Concerns about using bacteria, plants, and other organisms for these purposes should not be taken lightly, considering the environmental crisis we find ourselves in is now largely acknowledged to be due to the so-called Anthropocene, the period of human domination of earth. More specifically and accurately it has been attributed to the Capitalocene (Moore 2016) to highlight the differential control various groups of humans have had on the destruction of the earth and the differential impacts felt. Those in so-called third world or developing countries are bearing the brunt of environmental devastation.

Continuing to split those included or excluded in Science based on the seemingly new and more open criterion of passion instead of intelligence has not actually changed a division between "thinking" labor and "manual" labor. Some are deemed worthy of the passionate creative work of developing projects, and others are left to do the "automated" work that requires what is thought of as less valuable, repetitive, simpler technical skills. Further, those in the first creative, passionate, thinking category are included in democratic decision-making while others are left out. Outsourcing operates within a DIY biology lab according to an unquestioned adherence to capitalist values.

Other attempts at increasing democratic control over sciences and technologies are revealing. Some argue that computer geeks created a counterposition to intellectual property ownership through the open source movement; other activists and scholars have been critical of the ways that these supposedly apolitical actors unintentionally but no less dangerously have made way for further capitalist control through technology. Kelty (2008) has argued that a "recursive public" of geeks formed the open source movement, which acts as a checks and balances system to the power of government and big business to maintain democratic technologies; however, Jodi Dean (2010) has critiqued this analysis for placing these open source "geeks" outside these powers as a politically neutral force. Technology activists, such as the radio "pirates" whom Christina Dunbar-Hester (2014) studied, generally criticized the free software and open source communities for not being politicized. And, while open software proponents use a language of democracy and justice that seems to be a challenge to business as usual, Dean's assessment of the danger in this seeming neutrality is that "geek norms emerge, claim neutrality and appropriateness, and then retreat, leaving in their wake a pro-capitalist, entrepreneurial, and individualistic discourse of evaluation well suited for the extension and amplification of neoliberal governmentality" (2010, 22–24).

In the case of tinkering geeks of community synthetic biology labs, a similar pattern emerges. The introductory messaging was about democratic, open access to science. The end goal that emerges within a few years is producing profitable discoveries through scientific entrepreneurship. Initial discourses of neoliberal democratization open the door for capitalist projects that can reproduce familiar patterns of unjust distributions of access to knowledge and wealth. Without a shared anticapitalist politics through democratic means, individual economic freedom wins out.

Tinkerers seemed to mostly identify as antiracist and antisexist. Unlike the computer hacker culture that was described as not "family friendly," these labs believed in a more liberal, inclusive ethos. The porn has come down from the walls, and they are interested in talking about increasing the number of women and minorities doing science. When I first started looking at these spaces, I had assumed that DIY biology might reject the ideas in favor of increasing inclusion that were prevalent at the time in universities because they were not beholden to those policies.[10] Although they did not explicitly talk about diversity in their mission statements, these labs were not opposed to working on a more inclusive space—especially if there was grant money involved.

Even without financial incentive, there was a general assumption that all involved believed that more women and racial minorities in the space would be preferred and that explicit racism and sexism were to be avoided. This is not entirely surprising considering the popularity of a certain kind of antiracism and antisexism in corporate diversity that seems to mainly be used for marketing purposes. However, as I paid closer attention to the purposes and practices of the labs, I found this to be a typical liberal, antiracist, antisexist political position. The goals of science were limited to control over biological matter and entrepreneurial profit. The labs did not leave room to deeply interrogate Science itself as part of the production of these categories of difference. If only those who already have a passion for science are included, what kinds of true democratic experimentation can take place there?

The fields of synthetic biology and DIY biology emerged in the early twenty-first century into a world celebrated as postracial and postfeminist. This moment of emergence for synthetic biology matters as a colorblind and genderblind politics is interwoven into the figure of the tinkerer. The assumption that opening up a space to "everyone" will be good enough for democratization ignores the co-creation of the sciences with gendered racial capitalism. Science was integral to legitimizing a new global world order over the last centuries. Within this science-based order some were and still are identified as not or less than human. Unequal societal divisions mirror the prevailing scientific thought of the times.

Under gendered racial capitalism, the exact groupings and boundaries may change slightly but without challenging the intersecting system there will continue to be a logic that says some people are less deserving or do not need to be included due to some biological rationale, whether that be intelligence or passion. These exclusions continue to follow the gendered and racialized patterns that have developed over the last 500 years along with the global capitalist system.

CONCLUSION: A PROGRESSIVE RETURN NARRATIVE TO MAKE SCIENCE GREAT AGAIN

The tinkerer and DIY community biology labs fit into the history of the modern scientist, embodying a shift from intelligence to passion as a prerequisite for inclusion. That shift does not challenge science's objectivity nor what groups of people are included in or excluded from scientific practice. The passion for science ends up being a passion for inclusion in entrepreneurial capitalism.

The ideals of the new maker and tinkerer movements are not new or unique to this entrepreneurial turn but rather fit into the production of a particularly American white masculinity that embraces a supposedly hardworking underdog. What is a new twist to this moment/movement is the co-optation of racial and gender identity–based movements for rights at the same time as incorporation of a postracial and postfeminist colorblind, genderblind politics. The tinkerer as disenfranchised uses the rhetoric of being excluded that plays on the rights-based arguments of movements for social justice while not explicitly taking up any of these movements.

Alessandro Delfanti (2011) claims that synthetic biology ethics combines a "naïve narrative" about a return to pure scientific ethics that was based on providing objective scientific knowledge for the common good (before corporations ruined science) with a hacker's ethics that is antiauthoritarian to create a new "form of open science culture that not only embodies elements related to openness and sharing, but is rather a more complex recombination in which alongside these, other characteristics emerge: antibureaucratic rebellion, extreme informational metaphors, institutional critique, autonomy, independence, a radical refusal of external interferences and also of scientific institutions themselves, hedonism, the importance of being an underdog, and finally an intense relationship with the media" (53–54).

Reading this return narrative as "naïve" perhaps misses the importance of how this tinkerer ethics is gendered and racialized in particular ways, is attached to economic success, and is deeply American in its character. It is not simply a naive understanding of Science as great before; it is part of a dangerous narrative that ignores the history of Science as part of coloniality. The word "naïve" signals a kind of innocence similar to the way tinkerers narrate their interest in "exploration" and "entrepreneurship." This innocent reading of the field may leave tinkerers unaccountable and remove responsibility for the material-discursive moves they make. This is not to say that these are intentional moves. Rather, there is an ethical imperative for those of us interested in democratic science to pay attention to the ways we claim space and do science.

INTERLUDE 3

Learning the Limits of Ethical Debate

All students in the Graduate Division of Biological and Biomedical Sciences were required to attend the two-day in-person ethics training. Likely this was required due to National Institutes of Health (NIH) funding requirements. Most of us (perhaps all of us) were being funded at some point in our training through an NIH grant. One hundred of us, plus or minus a few, packed ourselves into a large lecture auditorium for the two-day training. Because it was mandatory, we had to sign in each day. After the first session, I decided I would leave; the next day, I showed up to sign my name then I left to do things that I deemed more important and more ethical.

When I was hired as a government ethicist after graduation and continued to move in bioethics circles, I found my limited ethics training to be telling of the state of science ethics training.

During that first training session, we were asked to spend a few moments writing down an ethical dilemma that we had encountered at some point in our first year of graduate school. Having recently become more politicized, I had encountered more vegetarians, vegans, and animal rights activists, and these folks had asked me questions about how I had decided to do animal research. At the time, I had not really thought about it as a choice.

I had been asked this same question once when I first began my adventures in neuroscience as an undergraduate student at a small liberal arts college. I was majoring in physics and math but had decided that neuroscience was more exciting. I saw it as more of an applied science: I could use math and physics while working on problems that had to do with humans—mental illness and health. I found the only neuroscience lab on campus and asked to work there. The principal investigator (PI) was extremely supportive and gave me a job and mentored me toward my graduate career in neuroscience. That lab studied the playfulness of rats, but I did not care what they were studying; my primary goal was to gain experience in a neuroscience lab for my graduate application. Then one day while I was walking

INTERLUDE 3

on campus with my mentor and primary academic advisor from physics—who was the only female faculty member and also a vegetarian—she asked me about my decision to do animal research. I put off the question, telling her that I did not care about it one way or another.

As I sat in the ethics training session, I was wrestling with this question again, now that it was frequently being asked of me by my political comrades. I was excited to write this dilemma out and submit it for discussion—and I was shocked when the ethical dilemmas selected for discussion did not include mine. Almost all of us in that room did animal research, so how could this not be an important issue to discuss? Instead, we spent most of the time talking about how to ethically determine who got credit for the work that we did, whether it was ethical to date each other or our professors, and other, similar topics that seemed to have to do more with professional ethics than specifically about biological sciences.

This is not to say those topics were not important. Yet how they were covered had more to do with the university's liability than actually discussing issues of labor, profit, sexism, and power.

Around that time I decided to stop doing nonhuman animal research, but it was hard to find a neuroscience lab that intentionally only worked with humans. In the lab where I ended up, I found that any time that I questioned nonhuman animal research my reasoning was unintelligible to my colleagues. Even though they knew I was vegan, the idea of a neuroscientist animal rights activist was an impossibility. Although I had found a way to keep to my own ethical commitments, there was never an opportunity to have dialogue with others about animal experimentation.

In a twist of amusing ethics anecdotes, I was given credit for my work on playfulness in rats, and this is now the first and most cited (up to this point) article on my curriculum vitae.[1] Is it coincidental that my academic travels have brought me to the subject of play once again?

As I discuss throughout *Labs of Our Own* and in other recent work, I now not only see issues like animal research as an important ethics issue to discuss but, more importantly, view ethics as a process, one that we must constantly be engaged in as we produce knowledge. Reading my first credited published paper (I am listed as fourth author), I now have so many questions about how we were defining play and its relationship to humans and to ideas of normality. In that work, juvenile play in rats was quantified based on counting nape contact and pins, and characterizing responses to these. This work claimed that studying play behavior in rats could help us to elucidate the genetic and cultural underpinnings of healthy play in human children. In that work and in the literature on play these sets of behaviors called "play" in rats are considered to be signs of creativity, learning, socialization, and risk-taking. The appropriate or normal amount of childhood play is associated with successful, healthy adulthoods whereas abnormal amounts are associated with neuropsychiatric disorders. Of course, what is normal and abnormal is determined based on the cultural assumptions of investigators and the dominant cultures at large.

In this next chapter, I focus on the ethical consequences of tinkerer communities. At the center of this work for me is asking what ethical questions and whose ethical questions are taken up and which ones are obscured and ignored.

CHAPTER 3

Becoming the Informed Public

Along with many researchers, I have been annoyed with the typical ethics processes that we must go through in the university. For most of us, we only come under ethical review if we are conducting human or animal research, at which point we must gain approval from our institutions' respective research boards. More than once I have had an institutional review board (IRB) application returned for not having the date/time prompt on an informed consent signature page in the format they wanted. I have felt both the irritation of what seems like a bureaucratic process of dotting i's and crossing t's correctly and a related concern for how the focus on IRB and informed consent limits deeper sustained ethical engagement.

Most researchers in synthetic biology do not have to go through a formal ethics review at their institutions because working with bacteria, fungi, and plants does not count as animal research. I, however, studying the scientists who conduct synthetic biology experiments in community laboratories did need to go through an IRB review of my project. We developed an informed consent form through this process which we shared with our informants before formally interviewing any of them. Before we began, we described our research and left room for questions. We asked if it would be okay to record the conversation for research purposes and gave them the consent form for review and if they agreed, their signature. We had left some forms with potential informants ahead of time so that they could look at it before we returned. Many of us are so used to signing documents that limit our legal rights and/or give us information that we cannot possibly process in the moment, that we simply accept and sign fairly quicky without much concern. Most of the do-it-yourself (DIY) biologists did not care much about the informed consent process. This made sense to us as the potential danger to any individual from being part of a cultural study of a public space seems to be nonexistent. Only one informant paused and wanted to talk more about it before we hit record.

The eventual interviewee was concerned not for their own well-being but more so for the community. They asked whether we would be comparing different lab

73

cultures. When we discussed further to understand the concern, we found out they wanted to make sure that our study did not cause a competition between different labs. They did not want us to report back that this lab is doing a good job and that one is doing a bad job. What we understood from this conversation was that they were quite concerned about the cohesion of the DIY biology community. In response, we said that we did not plan to do comparisons because we were more interested in the commonalities across the lab spaces. We did say that talking to us and the work we publish could possibly have an impact on how people think about community biology and may even shift our informants' practices or positions. After all, we do hope our research is relevant and impactful!

The concern for the broader image of the community and the desire to have a shared outward facing presence came up in other moments. For example, one day a lab member brought in copies of *BioCoder* to share with the group. *BioCoder* (2013–2018) was a free magazine primarily focused on DIY biology. The editors argued that this was the newsletter for the biological revolution that was about to radically change our worlds: "We're at the start of a revolution that will transform our lives as radically as the computer revolution of the 70s. The *biological revolution* will touch every aspect of our lives: food and health, certainly, but also art, recreation, law, business, and much more."[1] The lab members passed around copies of *BioCoder*, ready and excited to serve in the revolution. Like other revolutionary communities, the members sat around debating the contents of the magazine and DIY biology more broadly. A real sense of community and community identity was forming because of the intentional space that had been created for gathering.

The article of most interest was about the glowing plant experiments that had become perhaps the best-known output of DIY biology (Keulartz and van den Belt 2016). A conversation and consensus began to form, not around whether they should be able to do these experiments but on whether the glowing plant scientists were being thoughtless about the impact on the broader community. They were worried that the negative press that group was receiving threatened to shut them all down before they were able to really do much. There was a strong sense of community identification and a shared feeling that DIY biology was in a fragile early stage of development. DIY biologists appeared to be aware from the beginning that their public image was very important.

The two biggest ethical issues facing the field of synthetic biology can be characterized as concerns over (1) bioerror and (2) bioterror. *Bioerror* means an unintended consequence of genetic manipulation that leads to harm, and *bioterror* represents an intentional act of using biology as a weapon. As the democratic arm of the synthetic biology revolution, DIY biology added another layer to these concerns. Ethics conversations about DIY biology focused on whether the likelihood of bioerror and/or bioterror increased by adding amateurs to the mix (Kaebnick 2014). These potential dangers were to be weighed against DIY biology's promise, which was also framed as an ethical imperative to open science up to more people. This balancing act played out in labs, on websites, and in policy conversations.

Bioerror and Bioterror

To respond to concerns about safety and terrorism, community labs often make references to adhering to safety regulations set forth by the government on their websites. For example, BioCurious states that it is "a training center for biotechniques, with an emphasis on safety" and that their "biology lab functions at a Biosafety Level—1 level (BSL-1), which is equivalent to what you would find in a high school biology lab. We require all our members to undergo safety orientation."[2] Maintaining that community labs are safe because they are not really doing anything more than what you do in a high school biology class and at the same time arguing that something radically new will be created in these spaces may cause a tension. In the previous chapter, I uncovered a related tension around wanting everyone to be able to tinker with biological matter and the desire to create real profitable results. One limit to doing what might be considered more serious work is the need to prove that these unregulated amateur spaces are safe. The claim inherent in this line of argument is that the spaces do not need regulation.

Another response to the concern about safety and the potential need for regulation, oversight, and/or restrictions on biological tinkering was to create an internal code of ethics. One of the co-founders of DIYbio asked at an early think tank meeting for DIY biologists, "How do we build a positive culture around using technology and become good biocitizens?" (Tocchetti 2014). This was an important part of the conversations that led to the development of a code of ethics aimed at calming public fears through the media.[3]

Despite these pre-emptive ethical moves, government agents still wanted to discuss the possible dangers with DIY biologists. In June 2016 the National Academies of Science, Engineering, and Medicine (2017) held public meetings on "Future Biotechnology Products and Opportunities to Enhance the Capabilities of the Biotechnology Regulatory System." During the session on DIY biology, they asked about whether shipping genetic material to individuals should be regulated. In response, Ellen Jorgensen (one of the founders of Genspace) mentioned that not only is everyone already doing it but that she does not see this as an issue unless its "some hostile nation-state." Jorgensen's response suggests that there is not much concern for bioerror and that the focus should only be on bioterror. In contrast to the danger of "some hostile nation-state," she implies there is no problem with "someone in a dorm room receiving a beta galactosidase gene." Beta galactosidase is commonly used in genetic engineering as a marker for whether a genetic change was successful. In our bodies it is responsible for breaking down lactose, so it is quite abundant in nature as well as in the laboratory. It is unclear why this enzyme's gene would be more dangerous in the hands of someone from a hostile nation-state than a college student. Embedded in the bioerror/bioterror contrast is that *who* is manipulating genetic material matters.

We Are Proper

When Jorgensen refers to a hostile nation-state in a taken for granted/of course tone, she is positioning DIY biologists as "safe" through distancing from presumed terrorists: it is not dangerous for someone in a dorm room to be allowed to receive this shipment. Jorgensen's response is notable for illustrating how the DIY community is discursively produced through the boundaries it sets up, with the imagined college student (presumably white and not from a hostile nation-state) inside, and dangerous state (and nonstate) actors outside. Hostile presumably means hostile to the United States or the West more generally, thereby signaling a racialized other, the specter of danger, as the limit.[4]

Distancing from the criminal or terrorist was evident on some websites. For example, LA Biohackers explain that hackers are not bad: "Despite the common misconception in the media that 'hackers' are villains bent on stealing national secrets or vandalizing your Facebook profile, the term has a much more benevolent meaning amongst the people to whom the term is applied."[5] The DIY scientists define their counterculture identity, "hacker," against those who would threaten national security. Those who would use hacking for the nefarious goals of stealing national secrets are, of course, considered terrorists, not DIY scientists.

When I first visited the LA Biohackers website, there were several images showing what their lab space looked like under the subpage titled "Habitat." I was struck by an image that showed a white marker board with the words "NOT A METHLAB" written on it with an arrow pointing presumably to a lab area out of view.[6] Presumably it is meant as a joke, the phrase bringing attention to the possibility that lab materials could be used and/or assumed to be for purposes of drug-making. But also implied in this image is that those who would be involved or accused of operating meth labs are not the type of person who is included in the "anyone" for whom these labs are set up; it is funny and possible to write on the board because there is no way that these people in this space would be accused of such a thing.

Although methamphetamine users and those operating what is colloquially known as "meth labs" are predominantly white, male, and young (Weisheit and Wells 2010), DIY biology community labs are in urban environments, and the white, young, overeducated men who operate them are generally have advanced degrees. As much as DIY biologists rhetorically distance themselves from traditional biology, their proximity to it protects their class from association with illegal drug users or producers. This kind of language establishes both who is an acceptable DIY scientist and what the proper purpose of DIY labs is. Why, after all, are meth labs not considered DIY science spaces? What would it mean for such spaces to be included in the DIY category? Similarly, why is the hacker who steals national secrets not included in the definition of proper hackers?

The phrase "not a methlab" was also used by DIY scientists when preparing a document for FBI agents to explain that DIY laboratories were safe and DIY scientists were responsible people (Tocchetti and Aguiton 2015). In the labs that I visited, the relationship between the FBI, Homeland Security, and DIY biologists was evident from the flyers on the walls giving information about reporting suspicious people and the Homeland Security stickers on the laptops. Additionally, these labs have willingly worked with the FBI and Homeland Security to put on joint workshops.

Sara Tocchetti and Sara Aguiton (2015) documented what they argue is a mutually beneficial collaboration between the FBI and DIY biologists in which DIY biologists use the collaboration in shaping their image to the media to gain public acceptance. It also, however, creates an image of being bad boys—but not too bad—which is necessary for being positioned as what I call the "proper public." Meth labs and the popular narratives around meth labs have been played out in particular through the popular television show *Breaking Bad* (2008–2013), which gave viewers an opportunity to play tourist in the drug world of methamphetamine (Tzanelli and Yar 2016, 11) through a sympathetic white male who was struggling against a poor economy, increasing health costs, and stagnant wages for middle-class workers. In this way, meth labs have come to be a signifier of illegality; but also, through their proximity to whiteness, they indicate either an indicator of unfortunate circumstances or an ingenious approach to a system always against the "working man." Hence, the joke "not a methlab" both distances from illegal activity at the same time as bringing into view a deviant but somewhat understandable way that white middle America is responding to economic decline by taking matters into their own hands.

While not necessarily celebrated, through popular imagery such as *Breaking Bad* meth labs have become a metaphorical way of understanding the challenges facing this demographic and the necessity of sometimes using extralegal means to deal with their problems. This is where the acceptable bad boy as a sympathetic subject worthy of being the proper public comes into relief. This figure is similar to that of the hacker who is variously celebrated and feared (Turkle 2005). This has a lot to do with the hacker's race, class, and nationality. A white, U.S.-born teenager may be heralded as a prankster genius, but the fear of Russians hacking U.S. elections in 2016 has revived long-standing red-baiting fears of "the Russians" as the Others (foreigners) who will bring down democracy as the United States knows it.

The line that DIY biologists are straddling between criminality and acceptability may be finer than they realize at times. As I shared in the first chapter, DIY biologists reported being "abducted" by FBI agents. Some of the early DIY scientists I spoke to assumed they were on no-fly lists and extra security lists due to difficulties that they had flying. Being on no-fly lists and being shuttled around to undisclosed locations by FBI agents suggested a story of being on the edge of the law. However, the biologists surprisingly seemed to be less concerned with the kind of intimidation that seemed to take place than my research assistant and I, who had activist backgrounds, were. Based on the accounts of DIY biologists, we came to understand their cooperation as not simply chosen willingly. And while there was a lab agreement to work with law enforcement, there appeared to be a range of attitudes regarding their relationships with these agencies. In a blog about science policy and outreach on the journal *Nature*'s website, Rayna

Stamboliyska (2012) noted that there was disagreement in the DIY and synthetic biology communities about their relationships with law enforcement and the Department of Defense. Although there is a diversity of personal opinions and some debate within DIY communities about their relationship with the law, what emerges from these discourses is a portrayal of DIY biologists as straddling the line between acceptable and dangerous.

Community labs also distance themselves from traditional sciences, painting a rebellious, revolutionary image analogous to a popular narrative about the early days of personal computing. Yet, as in the story of computers (J. P. Kelly 2009), the DIY biology revolution has limits. Biohackers attempt to carve out space(s) between the elite scientific institutions that hinder their creativity and the criminal space of "the outlaw" (Kelty 2010). Morgan Meyer (2016) argues there are multiple disparate comparisons being made at the same time between biohackers and the figures of Steve Jobs, the Victorian gentleman, the terrorist, and the punk. Meyer argues that these analogies work somewhat independently. However, the space carved through the fluctuations between these various comparisons effectively creates the proper ethical subjects of DIY biology.

Moving between more and less respectable representations produces subjects that are proper citizens who have just enough edge to make them interesting and are more relatable to a diverse public. The figures of Steve Jobs and the Victorian gentleman are both of white, rich men with independence whereas the terrorist and the punk are countercultural subjects, racialized and classed as non-white and poor, who threaten the status quo in unpredictable ways. The ability for community labs to be defined more fluidly in the space between these analogies sets in motion ethical debates about the balance between freedom to innovate and the dangers of bioterrorism. In those debates, community lab members position themselves as different than foreign terrorists or criminals. That is, they aim to explicitly position themselves as "good" in contrast to dangerous actors. In the development of the DIYbio code of ethics, "good" was invoked explicitly along with biocitizen. The goal of becoming "good biocitizens" helps to define the DIYbio community as the "proper" public.

An image of unacceptability is positioned against the DIY scientist in order to show the limits of the latter's rebellion. One image is that of the illegal drug maker/dealer, another is the foreign terrorist. In broader ethics conversations about synthetic biology, the fear that this technology would get into the wrong hands is prevalent. The issue is that bioterrorists might be able to use synthetic biology technology to harm the West. In displaying the message "not a methlab" or offering that they are not terrorists, DIY biologists affirm that theirs are not the wrong hands. They are legitimate users of this technology and can be trusted.

Although terms such as "everyone" and "all of us" seem to code for a belief in a postracial society, evoking ideas of the terrorist, outlaw, or criminal have never been race or class neutral. These images help to define these spaces as safe, not only through the adherence to regulations but also through the racialized boundaries created through the distancing from antisocial actors and actions. In ethics debates, the first concern surrounding these new technologies and their regulation is bioterrorism, and therefore bioterrorists. That means that the formation of this community of tinkerers must be understood in the context of the ongoing "war against terror," which has been used as code for a white U.S.- and European-led war against "the Arab and Muslim world." The colorblind application of inclusion for all thus already excludes those not considered "us." This positioning allows this community to participate in ethical debates and work jointly against bioterror.

Bioerror and bioterror are racially coded so that bioterror comes from non-Western entities and bioerror is something that happens in Western territories. In fact, the community's national boundaries are such that there are zero DIYbio communities identified in Africa and only a handful in Asia and Latin America.[7] The majority who make up DIYbio are in the United States and Europe.

The racialized distancing from terrorists and criminals is a key way that the community argues that it can be trusted to not engage in bioterrorist activities. In discussions of bioerror, on the other hand, community labs propose to counteract this more innocent concern by using science literacy and education to form an *informed* public.

WE ARE INFORMED

DIY biologists do not simply advocate providing access to resources; they also demand access to the language of science. Several of the sites offer classes to members and non-members. These classes (sometimes called "workshops" and "training centers") are hands-on and teach scientific methods to those who attend. In describing the importance of their workshops, Open Wetlab explains that they "emphasize the principles of demystification and democratization," and this linking of democratization and demystification underscores the importance of science literacy for all. In another example, Biologigaragen states, "The aim is to give these groups and citizens a space where they, without prior knowledge, can learn about the biotechnology around them—how it functions and how to apply this knowledge and self-made equipment in their everyday life." Genspace has a similar explanation of their class offerings "to people with no prior lab training."[8] Each strongly emphasizes that this education must be accessible to those without previous experience, and that such access is part of their focus on democratization and increasing access to science.

Although the number of active community biology labs is small compared with the larger fields of biomedical and biotechnology and even synthetic biology specifically, the public-facing work that the founders participated in was significant and can be found far beyond their websites. They sought out press coverage, engaged in public policy conversations, penned op-eds and journal articles, and were studied by researchers. When I visited my first lab space, the DIY biologists sat down with me to hear about my research. I expected them to be excited about our shared interest in democratic sciences but thought they may be confused about what my research would involve. Instead, they expressed fatigue at how many social scientists and reporters had been interested in studying and interviewing them. To them, my research signaled a frustrating reality: more attention was being paid to the possibility of potential dangers of DIY biology than to the actual work being done in the lab spaces. In earlier days, they wanted press and wanted to share their revolutionary goals far and wide. At this point, they were not saying no to interviews or studies, but they were not jumping to make extra time for them either. They needed to address the doubts and fears from both the general public and the established biomedical institutions.

Shortly after opening, two of the founders of Genspace wrote a *Nature Medicine* op-ed directed at the biomedical field imploring them to embrace DIY biology (Jorgensen and Grushkin 2011). In the opinion piece, they argued that this "grassroots" movement of community biologists should not be judged on whether they can compete now with Big Biotech and university innovation; instead, they suggested, one of their key functions can be as "biomedicine's ideal ambassadors to the public" (411). "The community lab idea is an experiment, but an extremely important one. We don't know where it will lead, but even if the sole benefit is to enhance science literacy, that would be enough" (411). Contrary to the argument that DIY biology stands in opposition to traditional sciences, the reference to "biomedicine's ambassadors to the public" suggests that biomedicine and the public are distinct entities, and that DIY biology exists (or would like to) within biomedicine. Is "the public" those members of the public who show up to take classes and participate in the community bio labs? Or does "the public" refer to the broader Public, and do the labs (with some publics included) act as ambassador to larger public/societal concerns?

The op-ed suggests the former: "By providing outreach in the form of courses and workshops for nonscientists in subjects such as biotechnology, we make the latest medical advances more accessible, understandable and less threatening. It's harder to fear something one has actually done side by side with one's high-school-age son or daughter" (411). However, if the community biology lab acts as an ambassador for biomedicine to the members of the public who show up in the lab space, what happens when very few people participate? And if the labs are democratic spaces, should the participants be actively creating and changing what biomedicine and science means and does rather than becoming comfortable with an already developed field?

It would be a mischaracterization to say that the opinion piece only focuses on this ambassador role for citizen scientists. The authors also point to this moment being a great opportunity to embrace the public's interest in scientific inquiry. They cite members of the public sharing medical treatment results on social media as evidence of this democratic turn for science and medicine. Although the websites designed for messaging to a broader public often commented on how unfair it was to keep science out of the hands of ordinary people, when talking to the proper scientific community the DIY biologists focused more on the possibilities for public relations work.

Across different mediums, one commonality seemed to be toeing a line between the great radical potential of this experiment in democratic science and assuaging

TABLE 3.1

FEARS AND PROMISES OF DIY BIOLOGY SUMMARIZED FOR DIFFERENT STAKEHOLDER GROUPS

	Fears/dangers of DIY biology	*Promises/hopes of DIY biology*
The public	"Playing god," bioerror/bioterrorism	*Greater control and access to science*
Biomedical sciences	*Losing scientific authority,* bioerror/bioterrorism	Greater public acceptance of biomedical experimentation

Note: Seemingly opposing fears/promises italicized.

fears that this experiment was dangerous. Bioerror and bioterror were shared fears across professional and nonprofessional spheres. The public also evinced a more general concern about genetic manipulation as "playing god" (Dragojlovic and Einsiedel 2013), indicating not necessarily religious concerns about genetic manipulation but an ethical concern about scientists engaging in genetic engineering. The additional biomedical community fear that needed to be addressed was about losing authority. The promise for the public, however, was greater control over science and a decrease in traditional Science's authority over science. DIY biology was to expand scientific exploration beyond the hallowed halls of universities and the greed-driven labs of biotech corporations. For biomedical researchers, the promise seemed to be about making the public more accepting and loving of bioexperimentation. Therefore, these aims would seem to be in opposition (Table 3.1).

Both groups, although seemingly in conflict, prioritize science literacy. How science literacy is defined may be different in each case. The concept of scientific literacy for the public is historically entwined with public demands for more accountability and public criticism of new sciences and technologies (Giordano 2020). It is common in the sciences to argue that we must have a more educated public who will support science. When the public is skeptical of science, the response has generally been that "they" (the public) do not understand it properly. In contrast, science literacy for the general public has been pushed by social justice–oriented educators who believe in critical engagement with science (e.g., Roth and Barton 2004). Social justice advocates have argued for feminist or critical science education in which science is taught based on the needs of the community. They have argued for science education that will be useful in people's everyday lives. They have also argued that science literacy should not be conducted for more public buy-in but rather so that ordinary people can become part of critiquing and producing scientific knowledge themselves (Roth and Barton 2004). DIY community lab websites often invoke these social justice arguments for promoting science literacy as a way for people to have greater control over science.

This contradiction is ultimately connected with DIY biologists' efforts to carve a space between countercultural subjects and acceptable, safe, responsible entrepreneurs. In making a case for being "proper" and therefore not dangerous (causing bioterror) or reckless (causing bioerror), the community argues for opening up science to a respectable group of people and through training (science literacy) making this group safer at the same time as bringing them squarely into the accepting, science-loving public.

The language on community labs' websites includes modifiers such as "safe" combined with those promising greater accessibility, "supportive" and "affordably": "We [Genspace] work inside and outside of traditional settings, providing a *safe*, *supportive* environment for training and mentoring in biotechnology" and "BUGSS provides a space where individuals can *safely* and *affordably* investigate the living world."[9] Creating safe places signals safety from fear of bioerror and bioterrorism. "Affordably" and "supportive" suggest that these spaces will be more accessible. The Genspace statement describes what kind of science literacy will be taking place while the statement from BUGSS focuses on scientific discovery. In both cases, the role of science for discovery is strengthened. These constructions together might enforce biomedical professionals' confidence that the public will become more interested and welcoming of scientific experimentation.

The playfulness of DIY biology that I identified in the previous chapter is also important for the aim of making science more palatable to the general public. Showing scientists being playful and having fun makes them seem more relatable to the public and makes science itself seem more appealing.

The move to make science "fun" is not a new feature that has emerged with tinkerers. Cultivation of a stronger affective attachment to Science has been a concern for national science literacy projects since the mid-twentieth century when U.S. sciences began to rely on government funding more heavily. (Lewenstein 1992). In developing an argument for agential literacy, Karen Barad (2000) focuses on the field of physics pointing to a post-Sputnik promotion of "physics is phun" kinds of courses to bring U.S. citizens up to speed to compete with Soviet science. Barad points to the ways that the "boys just wanna have phu-un" approach had a lot to do with the popularity of Richard Feynman and his success in engaging more people in physics. Feynman's "love" for physics was evident from interviews while historians of science have suggested that his passion for science and understanding that more people were needed to solve problems made him not only a popular science figure but suggests that this embodied attitude itself was part of what made him a great scientist. Barad contrasts this approach with feminist attempts that might appear to have similar strategies to engaging with science. Feminist approaches highlight science literacy as one approach to give more people more control over their bodies and lives. Accountable and responsible science—not often thought of as fun (think feminist killjoy)—is a central tenet of this kind of democratization of science.

If the tinkerer is having fun playing directly with biological material and learning proper ways of being a scientist, what is the difference between the modern twentieth-century passionate scientist in a university lab and a DIY biologist in a community lab? Despite their substantial overlap with professional communities, the DIY biology community's case for being the Public rests largely on their being a space for "everyone" and "all of us."

WE ARE THE PUBLIC

The claim to be the public is made by relying on a binary formulation of Science/Public. By positioning community labs as not part of traditional sciences, they thereby become the public. Curiously, this is accomplished despite the fact that the organizers of such spaces often hold joint positions in traditional science laboratories, have been trained through these mechanisms, and make use of technologies and ideas that would not be possible without the relationship with such traditional spaces (Ikemoto 2017).

The content of these community science websites contains a strong undercurrent of antielitism, antiauthoritarianism, and antiestablishmentarianism. This is also evident in the tone of the websites—a tone that suggests a breaking free from norms and rules and protocols of traditional scientific communication styles. This tone matches a certain cyber age aesthetic in which a style of jeans and a T-shirt suggests comfort over impressing anyone—even at work. In asking for donations, for instance, BUGSS offers a way to participate: "And if you just want to throw money at us we'll take that as well."[10] The phrase "throwing money at us" strikes a familiar tone as contrasted with an example from a university website donation page or a big biotechnology company looking for investors. For example, the University of California–Davis College of Biological Sciences sought donations on its website in this way: "With a gift to the UC Davis College of Biological Sciences, you don't have to limit your philanthropic goals. You will be a key player in eradicating disease, feeding the hungry, rehabilitating the environment and developing sustainable energy sources, all while helping the next generation build a solid educational foundation for future scientific breakthroughs."[11] The tone on synthetic biology community lab websites and the use of "play" and "fun" mark a separation from stuffy science institutions that make science unfun and keep the public out.

Community labs claim the right to science on almost each of the websites that I analyzed. This follows a discourse put forth in Meredith Patterson's "A Biopunk Manifesto" (2010), in which she explicitly argues that everyone should have the "right" to biology (Keulartz and van den Belt 2016). In social justice contexts, rights discourse is used to combat exclusion (when a group is kept from enjoying the right in question). The organizers, who typically have access to science already, move back and forth between claiming inclusion for themselves and on behalf of the larger public.

Biospace argues that "access to knowledge shouldn't be limited to academia," indicating a broadly shared right to knowledge.[12] The lack of access to resources in the academy has been widely critiqued by feminists and others concerned about how access has been granted along gender, class, ability, and race lines.

The critique found through these DIY websites often lists educational, occupational, and sometimes economic barriers to access for all. The opportunity for a closer examination of how gatekeeping works is lost through the quick dismissal of academia as the culprit and tinkerers as the group deserving of inclusion. Although gatekeeping certainly takes place in academia, it is not the only place in society where it happens. Setting up shop outside the gates of academia does not necessarily mean that everyone would have equal access (particularly if those setting up the shop have the access required to move freely between both spaces).

With discourses of rights come discourses of our responsibility to exercise our rights. This imperative to participate is evident on several websites. For example, BioArt Laboratories explains that it "is an initiative that aims to make this (biotechnological) progress and its implications the collective responsibility of society."[13] BioCurious "is a completely volunteer run non-profit organization. . . . Joining the lab helps us continue to serve the community," and Biologigaragen urges "participation" by specifying that it "is a user driven project which relies on active participation."[14] Through this imperative scientific citizenship is defined, and the responsibility for inclusion is put back on those who participate or do not participate.

This rhetoric is common in debates about electoral politics in the United States, in which populations (and individuals) not exercising their right to vote are blamed for the lack of representation of progressive politics, thereby ignoring whether the electoral process itself is fair. In science, like in electoral politics, those who are opposed to the system itself or disenfranchised from the system should also have a voice because these systems impact everyone whether or not we participate in the prescribed ways.

The Co-emergence of Synthetic Biology, DIY Biology, and a New Ethics

To understand the impact of DIY biology communities on larger bioethics debates we must understand that the development of the DIY biology community is interrelated with shifts in academic genomic sciences. Although DIY biology does not account for the majority of synthetic biology work being done,[15] many accounts suggest a co-emergence of DIY biology[16] and synthetic biology in the early 2000s (Tocchetti and Aguiton 2015). If synthetic biology shares founding members with the DIY biology movement, does it also share foundational principles related to the ideal of more accessible, democratic sciences? Beyond the field itself, what role might ideas of more democratic sciences play in debates surrounding science ethics and public participation in science?

The field of synthetic biology emerged after the Genome Project and the National Human Genome Research Institute's Ethical, Legal and Social Implications Research Program (ELSI) in a period that was celebrated as postracial and postfeminist. This moment of emergence for synthetic biology matters for several reasons. For one, this science owes a debt to the genomic sciences of the past.

Christopher Kelty (2010) argues that Big Bio was and is necessary for this "rogue" science to prosper through the use of scientific and technological advances that make possible more efficient and affordable manipulation of genetic materials. At the same time, the ethics lessons learned from the interactions among scientists, ethicists, and nonacademic publics have also shaped the new ethics models being developed in tandem with this technoscientific endeavor.

Synthetic biology most commonly refers to an interdisciplinary field that aims to merge engineering and biology methods to create new technologies. The goal is to be able to synthesize segments of DNA to create biological entities with new functions or even create new organisms from scratch that will somewhat naturally reproduce within scientists' control and therefore produce a limitless source of raw materials. Although the promise of this kind of genetic engineering has elicited excitement and hope for some, it has also prompted fear and ethical concerns about playing God, creating life, and ensuring biosecurity. Synthetic biology has garnered so much attention that it was the topic of President Barack Obama's first Presidential Commission for the Study of Bioethical Issues.

Many synthetic biologists have argued against government regulations, proposing self-regulation instead. One attempt to formalize this self-regulation was made by DIYbio.org as they convened workshops to propose self-defined codes of ethics for the field in 2011.[17] A code of ethics for DIY biology was meant to minimize public concern (Eggleson 2014; Tocchetti 2014). The idea of self-regulation is not new (Weiner 2001). The 1975 Asilomar Conference on Recombinant DNA is cited as the first major example of scientists organizing to self-regulate around genetic technologies. The historian of science Charles Weiner (2001), after observing moves for self-regulation since the 1970s, argued strongly that this is among one of many maneuvers at play today in discouraging public involvement in scientific ethics decisions. Over the first decade of synthetic biology research, the media reported on it very rarely, and the coverage it did give acted to reassure the public that there is nothing new, nothing of immediate concern, and nothing dangerous happening in the field (Giordano and Chung 2018). Limited, ambiguous, and vague news reporting dissuades the public at large from learning about the field and becoming involved in decision-making.

The emergence of formal scientific ethics in the twentieth century has been, in part, about the relationship between the public and institutional sciences (Wilson 2012). Scientific research continues to rely heavily on government funding, so public support for projects remains important. Genetic research has come under particular ethical scrutiny. As scientists planned to map the human genome, anticipation of ethical concerns from the public prompted the development of ELSI (ethical, legal, social implications) programs alongside development of the technologies. However, many argued that ELSI programs remained ineffective due to their emphasis on evaluating downstream effects; they saw an opportunity to incorporate more collaborative, integrated models for ethics in synthetic biology (Rabinow and Bennett 2009). Jane Calvert and Paul Martin (2009), in arguing for synthetic biology to embrace a collaborative relationship with social scientists, point to the problematic ways that social scientists have at times been used to stand in for the public itself or have been thought of as translators or mediators between scientists and the public.

These new democratic sciences constitute ethical interventions in themselves, and their significance should be evaluated with this understanding in mind. Much as social scientists were made to stand in for the public (Calvert and Martin 2009), DIY biologists are now positioned as the public or "the people." This novel science should be understood within a long history of ethical struggles among scientists, bioethicists, policymakers, and publics.

At the same time as there are more calls for scientists to concern themselves with bioethics, there is also a continued push by scientists for increased science literacy in response to the lack of public support for certain science agendas. Scientists frequently bring up the need for an informed public in response to ethical deliberations about controversial sciences. So, as bioethicists have created more ethical practices for scientists, scientists have pushed for greater science literacy to produce a more informed public that will understand the need for science to move forward, assuming what has been called a deficit model of public understanding (Irwin and Wynne 1996).

Defining the Public and Participation

If public accountability and support are considered to be central to scientific ethics, then the public and its relationship to science need to be defined. Feminist and other science studies scholars have demonstrated over and over that the split between science and the public is a false binary. A wide range of scholars from differing disciplinary traditions have argued that scientific knowledge is always contingent on the historical and cultural context in which it is produced (Collins 1999; Haraway 1988; Harding 2004; Jasanoff 2004; Kuhn 1962). Although this point is not contested in science studies, attention to power and social justice is not universally accepted as integral to a clear contextual analysis of knowledge production (Campbell 2009; Pollock and Subramaniam 2016).

Feminist science studies scholars have maintained this point as central to our field. The study of formations of scientific knowledge must be grounded in an analysis of power, and, therefore, as researchers ourselves, we must be accountable and reflexive about power in our own knowledge production (Campbell 2009). The assumption that knowledge is always culturally contingent begins with the insight that scientific knowledge is produced by people and that these people (scientists) do not exist in vacuums. Therefore, the idea of the public as outside of science has always been complicated.

At the same time, there are segments of the public that have more and less access to science, and the benefits and harms of science are distributed disproportionately. The public that matters is formed through exclusions/inclusions in the proper demos that can be understood through the frame of gendered racial capitalism. A shift in framing may be needed from the *public* as a monolithic body, to *publics*

BECOMING THE INFORMED PUBLIC

TABLE 3.2

HISTORIC SHIFTS IN SCIENTIFIC OBJECTIVITY (TABLE 2.1) WITH ADDED
ROWS DEPICTING SHIFTS IN THE ROLE OF DIFFERENT "PUBLICS" IN THE
BIOLOGICAL SCIENCES OVER THE SAME TIME PERIODS

	TIME PERIOD (CENTURY)			
	17th—mid-19th	*Mid-19th— mid-20th*	*Mid-20th— late 20th*	*Late 20th— present*
Epistemic value	True to nature	Mechanical objectivity	Trained judgment	Objectivity contested
Scientist figure	Modest Witness		Expert	The Tinkerer
Attribute	Trustworthiness	Detachment	Intelligence	Passion
Source of funding	Wealthy patrons or self-funded		Government	Government/ Big Bio
Source of legitimacy	Theology (Christianity)/ public experiments	Objectivity/ distance from public	In the public interest (utilitarian)	Public participation

as discursively and materially forming, never self-evident or entirely stable. This task of determining the boundaries of communities within the larger public has proven no less difficult than describing the boundaries between the public and science (Fraser 1990; Warner 2002). Who, then, is the proper public to whom science should be accountable, how is it formed, and what is at stake in defining this public?

Who was included in the proper public and this public's relationship to science has been highly contested, featuring major shifts over time (Shapin 1991). These shifts are not neat moments when one set of ideas is completely exchanged for the next. As the definition of the Scientist changed, so too did the role and definition of the Public (Table 3.2).

As the biological sciences rose in importance in the seventeenth to nineteenth centuries, the role of the church remained so important that scientists made their case for legitimacy by claiming that their work furthered the goals of the church and did not conflict but rather amplified and illuminated God's work of creating the natural world (Shapin 1991). During this period, the modest witness also had to win approval of patrons to help fund their enterprise. To satisfy these religious and economic interests (likely intersecting), the modest witness presented his discoveries through public experimentation to these groups. The public who legitimated these experiments was made up typically of other wealthy men (gentlemen). Women, people of color, and workers (including technicians who performed actual experiments) were traditionally kept out of this public (Haraway 1997).

In the mid-nineteenth century, the move toward mechanical objectivity also meant a move toward a growing secularism in the sciences. The argument for pure objectivity meant that any influence from the public was considered to contaminate the results (Shapin 1991). Therefore, the sciences became further closed off. The problem of funding and concerns about the role of science after World War II led the sciences to be more heavily funded by Western governments and therefore more responsible to the public once again. This time the public was to learn through science literacy campaigns to trust and appreciate science. The promise was that the sciences would bring national prosperity and defense (through the focus on military science projects).

In the last couple of decades public participation and engagement efforts within scientific and government programs have proliferated in what Sheila Jasanoff (2003) has called the participatory turn. "The weakening cultural authority of science and interest in public oversight" have converged, creating both a need and demand for public participation (Kelly 2003, 341). Science studies scholars, scientists, and policymakers have debated, practiced, and analyzed techniques for creating the most democratic and legitimate sciences (e.g., Benjamin 2013; Campbell 2009; Collins and Evans 2002; Epstein 1996; Eubanks 2009; Jasanoff 2003; Kelly 2003; Lövbrand, Pielke, and Beck 2011; Reardon 2013; Stilgoe, Lock, and Wilsdon 2014; Wynne 2007). Some have asked questions about the legitimacy of different participation practices, criticizing forms of "invited participation" (Bogner 2012; Wynne 2007) where groups of public stakeholders are determined externally instead of self-organizing as a concerned community that focuses on the equitable distribution of power to create sciences that are more democratic (Campbell 2009; Eubanks 2009). Others have critically examined what, in this neoliberal moment, discourses of more just and participatory science do to/for ideas and practices of science, justice, and ethics (Benjamin 2013; Goven 2006; Reardon 2013). Ruha Benjamin's (2013) work on public participation in stem cell research debates in California illuminates the changing definitions of the people, with particular attention to which communities are mobilized as authentic public participants and how that has changed over the course of the political debates. That means that participatory biology is not only changing ethics debates but also creating new groupings of publics.

Feminist calls for more democratic sciences have been grounded in the insights of feminist science studies scholars who have combined science studies understandings that scientific knowledge is historically and culturally contingent with feminist epistemologies that argue for the importance of those marginalized in society creating knowledge (Harding 2004). Those calling for more democratic sciences have argued that the purpose of knowledge production and the methods for creating science must be based in investments in creating social justice (Douglas 2009; Jen 2015; Kourany 2010; Longino 1990; Minkler and Wallerstein 2011). Feminists' arguments for public participation in the form of more democratic sciences have tried to recenter the needs and knowledges of those publics traditionally marginalized from and by Science.[18] This is the discourse of disenfranchisement that is appropriated though the tinkerer and the tinkerer's community.

By arguing for more democratic science practices, DIY biologists claim to represent the public, which by common definition in the Science/Public split must be outside of traditional science. However, in one of the first examinations of the physical spaces created by community synthetic biology labs, Morgan Meyer (2013b) argues that the boundaries between traditional (university) science and DIY science are much more permeable than one might think, based on the declarations of DIY biologists. Similarly, by analyzing counterculture images such as the hacker and the outlaw versus the criminal, Kelty (2010) questions whether DIY biology marks a radical departure from traditional sciences or is instead "simply a leaky boundary between what used to be elite science and something slightly less elite" (6). These scholars and others raise questions of who gets to do science, who can be a scientist, and how the public is defined (Jen 2015; Kelty 2010; Meyer 2013). The way DIY biologists in particular rhetorically position themselves against traditional science has implications not only for fairness and inclusion but also for scientific ethics and public accountability.

As postcolonial, decolonial, and critical race studies scholars have pointed out, democracy and inclusion politics are not necessarily enacted to create more justice. Recently the United States has fought wars in Iraq and Afghanistan in the name of spreading democracy and creating gender equality. A belief in a postfeminist, postracial Western society has been used to silence critiques of continued racism and sexism as well as used to separate the West from "the rest" through a progressive narrative of democracy and equality. These ideas were more broadly invoked in the larger "war on terror" where the supposedly democratic and feminist West has used these ideals to wage war on majority Muslim countries. For example, and quite disturbingly, this rhetoric was even used to argue in defense of U.S. war waged supposedly on behalf of oppressed women in Afghanistan and Iraq that has killed and displaced many women (along with men and those of other genders). Not surprisingly perhaps, the formally connected DIY biology community includes almost no labs in majority Muslim countries.[19]

The foundation of liberal democracy itself relied on colonization and othering through the production of rights for certain classes of people (Lowe 2015). Working within these systems, disenfranchised classes of people have claimed rights as a matter of social justice. Thus, claims for democracy, rights, and inclusion are socially and historically situated, and they cannot be judged out of context. As a belief in democracy and inclusion for all is invoked as evidence in and of itself of a better, more just way of doing science, we must examine these calls for democratic science more closely because of the varied ways that democracy is and has been used as justification for a variety of inequitable social arrangements (gendered racial capitalism).

BECOMING THE PROPER INFORMED PUBLIC AND ITS IMPLICATIONS FOR SCIENCE AND ETHICS

The community DIY biology labs that I have studied stake the claim not only of being part of the public (as opposed to institutional science) but, more specifically, the proper informed public. Based on the large percentage of DIY biologists who are members of traditional scientific institutions, Meyer argues that DIY biology is not an "established 'amateur science' but rather a 'promised' amateur science, a citizen science 'in the making'" (2013, 123). Publics are discursive formations (Warner 2002), and the rhetoric of democratic science produces a specific public. A collective identity, tinkerers, is formed as an identity that should have rights (access to science)—not in the sense of trying to gain legal protection based on the tinkerer identity, but rather that the most understandable and acceptable social justice discourse (in the West) of inclusion and rights is evoked in the forming of this community as disenfranchised. The use of universals such as "everyone," "all of us," and the use of "we" in opposition to traditional sciences that keep people out creates the sense of a public that transcends all lines of difference (Tocchetti 2014).

At the same time as they are established as a public in opposition to institutional science, the language on the DIY biology community's websites makes it clear that they are good, law-abiding rebels. The sense that DIY biologists form a *proper* (as in respectable) public is connected to their presentation as being the most appropriate (proper) speakers for the public, based on their tinkerer identity. They are liberal subjects who are deserving of rights; they are proper citizens who are demanding inclusion and the right to do science. Through this careful construction, tinkerers formulate a demand for rights that is just rebellious enough to appeal to a sense of justice yet still nonthreatening to the (neo)liberal order from which (institutional, traditional) science garners its legitimacy in the first place. To be considered a legitimate public, they must appeal not only to the public at large but also to scientists. The appeal to scientists occurs through the promotion of science literacy and education, which demonstrates that the public will be informed (e.g., Jorgensen and Grushkin 2011).

What does the production of this proper informed public do? This framing suggests not necessarily a radical departure from the way science is done but rather an ethics intervention. Some websites mention the desire to engage directly in ethical discussions. For example, Genspace says, "The best way to inform the dialogue about 21st century science is to have the stakeholders understand it from a hands-on perspective."[20] Statements like Genspace's—that those making decisions (stakeholders) should have "hands-on" experience—position these informed stakeholders to have a unique claim to being the proper public. In the tradition of liberal democracy there is an imperative to join this proper public if you would like to have a say.

Although some biohackers promote the idea of a "do-ocracy" where the public has more hands-on control of biology (Wohlsen 2012, 38), there is also a danger in assuming that only those who are "doing" DIY biology should have a say in the kind of science they produce (Keulartz and van den Belt 2016). After all, the production of knowledge has never only impacted those producing it. As I showed in chapter 1, there is a specific type of person who belongs in tinkering spaces; one requisite is that the person is passionate about science. By default of wanting to be included, the passionate tinkerer agrees with a set of assumptions and values about the ethical value of synthetic biology and genomic science. Therefore, the terms and boundaries of the ethics debates are prescribed in certain ways.

The logical end result of tinkering is assumed to be entrepreneurial success. Therefore, the public is largely divided along racial and gendered lines into capitalist subjects, and those who wish to pursue democratic sciences as a challenge to gendered racial capitalism itself are rendered illegible. Those questioning the dominant role of genetics are sidelined through these criteria as well as those at the intersection of indigenous land sovereignty and environmentalism who are concerned about the use of biological resources for capitalist gain.

More Science = Better World

The claim that science can make the world better recurs across the DIY lab websites as an unchallenged assumption. For example, BioCurious encourages membership by linking the idea of more science in the world to a better world: "Become a member to make the world a better, more sciency place!"[21] In fields such as nanotechnology, scientists believe strongly that science benefits society, thereby closing off conversations about broader ethics (Bassett 2012). The assumption in this case is paired with a claim that doing science in a more transparent and democratic fashion will create even better results: science does good for the world, so by having more science and more ideas there will be more progress and innovation, leading to an even better world.

Although linking democratic science to better science has also been a claim of feminist scientists, the understanding of the link is quite different. In these new DIY science communities, the idea of democratic science seems to mean anyone can give input and practice science. The theory behind why this is better is that there will be more minds working on the project. Here quantity matters. Based on interviews with many synthetic biologists, including Drew Endy, Sophia Roosth (2013) found that there is a common belief in the field that their model of openness will lead to better biology. Feminist calls for more democratic science for better science begin from an assumption that science is not necessarily good (nor bad) but rather serves the interests of those who produce it. Therefore, when feminists use the modifier "better" it indicates a hope that science will change for the better by putting science into the hands of marginalized groups because serving their interests would create greater social justice. By contrast, it appears that DIY science communities begin with a vision of science as unquestionably good, which means "better" modifies it to an even more awesome version.

As the BUGSS site puts it, "We have thermocyclers and centrifuges and incubators and all the other tools you'll need to explore this world." Similarly, BOSLab's (formerly BOSSLab) website states, "The mission of BosLab is to open scientific exploration to all & to help solve problems we face on Earth."[22] These statements suggest we should have a universal investment in the claim that science is what is needed to solve our problems. There is an assumption that we are in it together as far as "the problems we face on Earth." This assumes that there is consensus on what the problems/questions are in the first place. It then puts forth a proposition that having access to scientific equipment and laboratory space will produce solutions to those problems. BOSLab is representative of an overall linking of democratic science to increased public good through its linking of making science accessible "to all" and solving our world's problems.

A later addition to DIY biology claims to create a better world for everyone is Genspace's spinoff project named "Biotech without Borders." Biotech without Borders is a nonprofit using the same space as Genspace and with apparently the same goals and mission. Its history begins, "Biotech Without Borders was founded in 2017 by people passionate about the potential for DNA science to change the world for the better." And its mission states, "Biotech Without Borders is a nonprofit 501(c)(3) public charity dedicated to democratizing the practice of biotechnology for useful and peaceful purposes in order to benefit humankind and the planet."[23]

The invocation of the "Doctors without Borders" name highlights again the appropriation of social justice rhetoric by a field that seems most interested in entrepreneurship. Doctors without Borders provides humanitarian aid throughout the world, particularly in areas of war, thereby sometimes treating people who are being harmed by Western neocolonial wars and practices. The field of biotechnology already operates without borders in many ways. Biotechnology is dominated by multinational corporations who make profit in a globalized world. While the movement of people across borders is highly controlled, capital has flowed in global networks since European colonization of the fifteenth century. Meanwhile, the idea of the "proper" biohacker is again enforced through the citing of "peaceful purposes," once again distancing DIY biologists from "terrorists."

CONCLUSIONS

A common claim across democratic synthetic biology websites is that science belongs to everyone. This is posed as a basic right. The BioBricks Foundation spelled this out, stating, "We believe fundamental scientific knowledge belongs to all of us."[24] The problem they pose and seek to correct through their work is that science has been kept out of the hands of ordinary people. DIY biologists' identity claims position them as the public (despite close connections with traditional sciences) and practitioners of democratic science. A "proper informed public" is formed by positioning lab members as outside traditional science but within acceptable social standards (in part by setting up racialized boundaries) and focusing on scientific literacy. The assumption that an entrepreneurially driven science will be able to solve societal problems is embedded in the formation of this particular public. This boundary crossing between Science and the Public and assumptions about the value of democratic sciences has a concerning impact on science, ethics, and public participation debates.

The focus of ethics debates has been whether tinkerers should be permitted to conduct science outside of traditional settings. This obscures other ethical debates about social justice related to the kind of knowledge being produced and its societal impact. For example, the issue of where the raw materials for the proposed products are taken from has been largely ignored, except by certain environmental activists (e.g., the ETC Group—Action Group on Erosion, Technology, and Concentration). There is an appropriation of traditional social justice rhetoric of inclusion and democracy that is mobilized to show that DIY community scientists are disenfranchised.

The tinkerer, however, as a universalized yet disenfranchised subject, undefined by race or gender, is positioned as the proper member of the public—one who deserves a chance to make money through entrepreneurial science. This explicit use of colorblind and genderblind politics ironically produces gendered and racialized forms of power in ethics debates. Juxtaposing All Lives Matter and Black Lives Matter, for instance, demonstrates that trying to universalize the message means ignoring the material differences in subject position in an unequal society, thereby exacerbating inequality. The substitution of tinkerers for the disenfranchised subject defines the terms for legitimate bioethical debate on issues of fairness and social justice. Through this intervention constraints are placed on larger ethical debates.

The claim to be the proper public potentially delegitimizes and cuts off other public concerns about the ethics of genomic research (and other ways of knowing or seeking a better world). If science ethics is about being accountable to the public, and the boundaries of science and public are merged in this case, then there is no need for external ethics because these spaces represent the public itself. Therefore, this group determines ethical concerns. The danger is that this potential redefining of the boundaries between the public and science is done in such a way that science-at-large may end up being *less* accountable while claiming more accountability/access.

Gendered racial capitalism has, over hundreds of years, maintained entire geographic locations (non-Western) as Othered as a way to distinguish those included from political/economic/scientific control. This process of Othering of non-Westerners means (1) that some biological tinkerers may not be intelligible as legitimate tinkerers (instead as bioterrorists) in non-Western territories and also (2) the extraction of labor and other resources can be conducted without concern for indigenous or local resistance (limiting questions of ethics and who can be legitimate stakeholders) from non-Western lands.

The lab rhetoric of the synthetic biology community mobilizes the idea of a universal subject—through language of access for everyone—at the same time as disenfranchisement discourse is used—pushing for the inclusion of tinkerers as a particular identity. This demonstrates the flexibility and dangers of neoliberalism in appropriating the language of democracy and the language of disenfranchisement. Gendered racial capitalism often disavows its continued disproportionate impact on women and people of color in order to continue these divisions (Hong 2015a).

The supposed gender- and race-neutral definitions of tinkerers easily reinforce capitalists (still overrepresented by white men) as proper subjects.

As tinkerers become the disenfranchised subject in need of a "lab of their own," the focus of ethics debates is redirected to the rights of the tinkerers, which not only might mean a lack of focus on traditionally disenfranchised groups but also may define the boundaries of the ethics debates around rights and access to science. The primary concern for justice or fairness in these debates is framed around the rights of tinkerers to freely innovate. In the case of genomic research, Reardon (2013) found that the call for justice and more democratic science created a shift from concerns about scientists harming the public to concerns about institutions hurting the public. Jenny Reardon suggests that the move to democratize and create justice through inclusion in genomic data collection and access to one's own data have replaced ethical deliberations over genomic research that were seen as cumbersome and limiting scientific progress. She cautions that these moves obscure broader ethical concerns about the choice to do genomic research in the first place, especially considering indigenous activists' concerns, both about more pressing social issues that are impacting their health and about not receiving adequate attention and funding. In the case of DIY biology, we must be attentive to what questions are moved to the front and which ones are obscured by calls for democratic DIY labs. DIY biologists claim an ethical position in opposition to traditional institutions of exclusion overshadowing other ethical questions.

We are being offered a false sense of choice between institutional sciences and democratic sciences. Wendy Brown (2015) argues that the way certain forms of inclusion and participation in governance decisions have been incorporated without any real "capacity to decide fundamental values and directions . . . cannot be said to be democratic any more than providing a death row inmate with choices about the method of execution offers the inmate freedom" (128). She argues that this kind of appropriation of democracy represents the "language of democracy used against the demos" (128). Efforts to democratize science projects within the academy or government (such as public engagement) have been criticized along the same lines (Tutton 2007). DIY biology's use of democratic inclusion—in this case into the institution of science—is superficial. Public engagement efforts are used simply to legitimize research, and they do not work to change or question research questions. A self-organized democratic movement (such as DIY biology) appropriates in the name of the demos without changing basic epistemological assumptions.

An ethical imperative to advance scientific research for the public good becomes a basic assumption put forth by DIY biology labs. This is echoed in major policy reports. The Presidential Commission for the Study of Bioethical Issues (2010) report on synthetic biology begins (and ends) with the assumption that scientific progress itself is an ethical position. The idea that community laboratories may advance science for the common good is suggested in part through a narrative of returning science to a purer form, before it was corrupted by big business and bureaucracy (though many have argued that science was never pure or objective:

e.g., see Delfanti 2011; Epstein 1996; Harding 2004). While *tinkerers* in synthetic biology community labs evoke social justice language and ideas for the rights of citizen scientists, the institution of Science remains good for creating a better world. We should be alert to this superficial appropriation of social justice rhetoric that ultimately excludes social justice claimants who might challenge the neutrality of science itself as part of making a better world.

INTERLUDE 4

Nerd Masculinity

As a white, queer, masculine, assigned-female-at-birth feminist who had trained as a scientist, I chose to pursue a fun, overemphasized, nerdy, masculine professional presentation at my first job after graduate school. However, the options for outrageous professional men's clothing are quite limited. This led me to the now uncommon and decidedly unfashionable—but well-recognized—pocket protector as a key wardrobe feature.

This strange choice seemed to make it clearer in the overwhelmingly straight environment of the Centers for Disease Control and Prevention (CDC) that my gender failure was quite intentional and that I was aware that I was wearing men's clothing. I had clearly not stumbled into a pair of men's trousers and an ill-fitting men's dress shirt by accident at the department store. The pocket protector provided a space for colleagues to ask me about my clothing choice in an acceptable way; by focusing on it, they could start the conversation, and I had an opportunity to control the conversation.

Typically it began with a confused look as they stared at the flap of the item in my pocket. The awkwardness in those moments made me feel content and much more powerful than the moments of staring that I would have received without this item. The appropriation of nerd masculinity typically reserved for white males has its potential and its limitations for challenging the gendered and raced gatekeeping of professional, technical worlds (Eglash 2002).

On one of the first days that I wore my pocket protector, one of my supervisors entered my office, gave the pocket protector a brief double take, then launched into the work-related question that had prompted her entering. After about ten minutes we wrapped up the work discussion; then she tilted her head and asked me about the pocket protector. (The most common question I am asked is, "Is that a pocket protector?" This translates to "Why are you wearing a pocket protector?" The anachronism coupled with the genderf-cking visual generally causes just the right amount of confusion to satisfy me.) I answered her, "Why yes, it *is* a pocket protector." She then asked if it protected my pocket from pens leaking, and I

INTERLUDE 4

explained that it was more of a fashion statement than utilitarian—but I did not seem to have many problems with pens leaking, so I supposed that meant it served its intended purpose quite well. Why was she interested? I asked her. Was she having a pen-leaking problem of her own?

My supervisor had already retired from her government position and was now working as a consultant. She had had a successful career at the CDC starting in the 1980s, and she was definitely earning a six-figure salary. So this was why I was shocked by what she said next: she lamented that she could not convince her husband to wear pocket protectors—cleaning the stains out of his shirts was really difficult. I was thrown off and became the confused one while trying to take in this new information: not only was she laundering his clothes, but he refused to take any steps to make this task easier for her or clean the ink off his own shirts.

How was it that a successful female scientist had this stereotypical 1950s-U.S.-housewife problem? This sort of disconnect between inclusion and systemic transformation at home or in the workplace is largely a known phenomenon. For example, studies continue to show that, despite working away from the home, women are responsible for more of the childcare and household work in their partnerships with men (Horne et al. 2018). This division has become even more obvious during the COVID-19 pandemic (Chotiner 2020; Kisner 2021).

In considering the domain of Science, feminist science studies scholars have struggled with the question of how to think about this disconnect between increasing the number of women and people from underrepresented racial groups in the sciences and changing the kinds of knowledge produced (Hammonds and Subramaniam 2003). This question of how "who does science" is related to "what kind of scientific questions are asked" and "what results are reached" and has prompted many feminists to come to the field of feminist science studies. In the next chapter, I examine how feminist scientists are politicized and how we might use this politicization and privilege as scientifically trained individuals to change the way that scientific knowledge is produced.

Does our success at appropriating a nerd identity through our science credentialing ultimately challenge the racialized and gendered ordering? Or do we end up strengthening the systems of exclusion?

CHAPTER 4

Feminist Labs of Our Own in Academia?

I was midcompletion of my PhD in neuroscience when I first learned about feminist science studies. I began my PhD in August 2001 shortly before September 11, 2001. Although my politicization no doubt had multiple earlier sprouts that were waiting to fully root, that moment was what began a rapid and deepening political radicalization. My first introduction to the field was through reading Anne Fausto-Sterling's *Sexing the Body* (2000). A queer feminist intersex activist who was shocked that I did not know much about the science of sex and gender lent the text to me. This book and the lessons from feminists around me left me wanting to leave the sciences altogether. But it was also ultimately activists who convinced me to stay and complete my PhD. Activists from the now defunct Committee on Women, Population, and the Environment taught me about the history of science and eugenics at a teach-in, and that really sealed the deal for me. I was not sure that science was something in which I wanted to be involved. But those same activists asked later if I could do scientific research on a question about birth control. Through conversations with them, I realized that there was a desire for more democratic, social justice–based science research. It was these conversations that made me decide to try to practice a feminist science, and that led me to the pursuit of the creation of a feminist science shop.

In 2005, I began to change my dissertation work to fall in line with my political objectives. I was able to interpret the results of my PhD work through feminist science studies scholarship and include this in my dissertation. While my committee became interested in my historical contextualization of my project and found the feminist science studies literature that I drew on interesting, ultimately no one was willing to offer me funding or mentorship support (these two things are related no doubt) after I completed my PhD to ask different (feminist) questions within Science proper (Giordano 2014).

After completing my PhD in neuroscience, I was committed to starting my next project with feminist science studies insights instead of adding on a downstream analysis. My goal was to conduct a basic science research project that was created together with the communities most impacted by the research. That meant people with disabilities that affected motor systems—people who had strokes and/or spinal cord injuries. During the most promising interview with a principal investigator who ran a multi-million-dollar laboratory devoted to walking for people with quadriplegia, I pitched the idea of a community-based project that would begin by developing questions for research with people with quadriplegia. Although the principal investigator agreed with the points I was making or at least understood them (as indicated by head nodding), after I was finished and asked if we could work together, the answer was an unequivocal no. He said that there had been a recent study conducted and published about what research questions people with quadriplegia were most interested in already. The results from that study showed that walking was not most people's number one priority. I sat shocked, looking around at all the expensive equipment for walking trials. He went on to explain that the problem came down to funding—money. The funders still saw walking as the most important goal, and that is where the research money was coming from to support his work (Giordano 2014).

The hyperfocus on walking as part of a curative politics of disability has been critiqued by many disability activists. For example, Eli Clare (Clare 2017) has critiqued Christopher Reeve (an actor famous for his role as Superman who became injured after a horse-riding accident) for his embrace of ableist notions of cure that placed walking at the top of his list. Clare explains that Reeve is not alone or the originator of these ideas but rather became a popular spokesperson for them, thereby increasing the chances that funding patterns would continue to focus on the eradication of disability instead of a more complex social justice-oriented approach. This is a cyclical problem where the kind of research/knowledge produced further strengthens ideas of what is normal and desirable, giving cultural assumptions a supposedly naturalized truth through association to science. After a few other attempts to convince principal investigators to support me in my community-based motor research project, I ended up leaving science proper (for ethics at first and finally feminist studies). I remained, however, in search of a way to produce relevant research for our communities' needs through feminist science labs of our own.

Although other feminist studies scholars who were trained as scientists each have had their own trajectory, there are some aspects of our stories that seem to be common among us. Many of us have struggled with attempts to continue in the sciences, in laboratories sometimes, while developing a more critical feminist analysis of science and the larger world. One way that we express a continued desire to do science as opposed to critiquing science is through dreams of creating feminist laboratories in the future. Some of us have met each other and dreamt together about building a feminist lab; others have written about the need for creating laboratories in women's studies (Subramaniam 2005, 2009). This has been a call for women's studies to "reorient" itself to scientific knowledge production, owning knowledge production in a way that does not leave science as walled-off territory. This is a suggestion about how to bridge the questions of representation of women

(women *in* science) and critiques of scientific knowledge itself. In early feminist science studies literature, this gap between the two was often problematized as a missed opportunity to really change science (Longino and Hammonds 1990; Subramaniam 2009).

The focus on "women in science" with a goal of increasing the number of women in the sciences often universalizes the category of women and ignores how class, race, and disability can continue to structure inclusion and exclusion. Although there has been an increase in white able-bodied women in the sciences, Black, Indigenous, and Latinx people across genders have continued to be systematically excluded across genders. Therefore, even those of us who left the sciences for political reasons or feelings of gendered exclusion often represent dominant racial groups in the sciences.

The problem of looking at women as a universal category in the case of representation in the sciences has also leaked into feminist science studies, where too much of the literature has focused on assumptions about gender differences in science studies while often missing out on how gender is always a racialized category. Therefore, we have well-meaning and often well-written rebukes on scientific claims that discuss gender as if it can be discussed outside of race.

Feminist science studies may benefit from a deeper engagement with feminist theories on intersectionality or other concepts that help us to see how categories of difference are co-produced, intermingled, contingent, and also specific (Subramaniam 2009). Further, by looking outside the academy we can understand how people who have been kept out of science proper have had their work appropriated at times and at other times have produced knowledges counter to prevailing scientific truths without the need for validation from Science or the university (Conner 2009; Rusert 2017). Identifying the multiplicity of ways that people participate in sciences may allow us to recognize the singularity of what counts as success in traditional STEM (science, technology, engineering, and mathematics) fields and show us ways to challenge not only these fields but systems of white supremacy and coloniality that they uphold (Slaton 2021). My interest here is to think carefully about the institutional and epistemological power of Science and make more room for resistant feminist sciences or antiscience approaches.

Put simply, the work of these chapters is to look at a community I call home, the feminist academy, and ask how our approaches to liberatory work might support and challenge the root problems. The Gesturing Towards Decolonial Futures collective provides a useful diagramming of various approaches to taking down what they call "the house modernity built" (Stein et al. 2020, 45). This sets up the problem as the structures built by and upon the colonial/modern divide, resonating with the arguments I have made throughout the book about the importance of this 500-year progressive difference with special attention to the Sciences. On one side they offer soft-reform and radical-reform spaces that they argue provide life support for modernity even as these approaches aim to correct injustices. On the other side, they offer a beyond-reform space made up of approaches such as "walking out," "hacking," and "hospicing" under the metaphor of palliative care for modernity.

Their suggestions, as mine, are not about supposing we can predetermine the correct path to take down all injustice. Instead, I draw on this framework and other queer, decolonial concepts to ask about both our intentions and impact of working from within the academic wing of modernity's house.

IS IT TIME TO MOVE "BEYOND CRITIQUE"?

The progressive narrative of feminist science studies suggests that it is time to move beyond critique. Current manifestations of this progress narrative appear in new materialist literature. However, this impetus may not be so new. Although the newness of new materialist feminists engaging with science, biology, or the body has been challenged (Irni 2013; Willey 2016a, 2016b), the call to engage is not new either. Almost twenty years ago, the co-editors of *Feminist Science Studies: A New Generation* explained in their introduction why they decided to use feminist science studies instead of "feminist critiques of science": "We along with many scholars, have come to prefer *feminist science studies* because the field does more than critique. It allows for progressive, positive readings of science, and of reconstructions of science consistent with feminist theories, ideals, and visions. We envision a future in which feminism and science are no longer oxymoronic practices but instead partners in creating a more equitable and just world" (Mayberry, Subramaniam, and Weasel 2001, 10). Here although critique is not discounted as obsolete in the field, engagement with science is seen as distinct from critique, and there is a hope that the conflict between science and feminism will disappear to create a harmonious pairing.

Going back another fifteen years we see a somewhat similar vision for the future. Bonnie Spanier reflects on a decade of women's studies in the academy and its impact on the natural sciences: "The blossoming of feminist critiques of the natural sciences and mathematics has just begun and will lead us to visions and immediate ideas about what feminist science can be" (1986, 71). In this earlier articulation, critique is quite central to the creation of a feminist science, yet there is still a progressive hope to move forward toward a new engagement with science.

What does it mean to work to make feminism and science "partners" or to dream of a feminist science? The definitions of feminism and science as well as feminist science must be carefully analyzed to understand what kind of larger political commitments are at stake. How we narrate our goals for a different relationship between feminism and science will certainly shape our feminist science practices. Through our dreaming and storytelling, we often express these relationships through affective attachments.

SCIENCE DEFECTORS

Based on the literature and personal conversations with other feminists, the metaphor of defection is not a popular way to describe our relationship to science.[1] It is one I began using as I started this project. Is defection an apt metaphor for feminists who have moved from the natural and physical sciences to women's studies programs? A defector switches sides due to an ideological difference with their political party (often nation-state).[2] The act of defection makes it nearly impossible to return to your original home due to the mark of treason. In a complementary fashion, there is an abundance of excitement on your welcoming new side, marking their sense of righteousness that someone has had a change of mind. It is more exciting to win someone over who was already indoctrinated by an opposing ideology than to win over someone who did not begin holding any strong opinions. At the same time, one who defects is not quite comfortable in the new home. There are things about the old home, which cannot be returned to, that are missed. This might come in waves of homesickness. Also, if one defects once, there is always a bit of distrust in the new home—you may be the type of person who could change your mind again—as well as fear that some old allegiances may remain.

How does one decide to defect from the sciences to women's studies? What longings for scientific homes remain for defectors? These two questions bring us back to the topic of this book—labs of our own. We left the sciences because the ideology of objective, rational, unattached knowledges did not serve our feminist needs. We could not become the knowledge producers we wanted to be in those spaces. And most of us tried. Our autoethnographic descriptions of attempts to continue to do scientific work reveal severe limitations to doing science differently within traditional scientific programs (e.g., Giordano 2014; Roy 2004). But we long for our experiments. As defectors often long for the camaraderie they left behind and their homes, we long for the kind of collaborative work science offered, the literal hands-on work and our lab spaces (although not all of us worked in wet-labs).

We left our labs of milk and honey for the promised land of politically conscious feminism. And when we got there, we missed something about doing science. But what is it? Is it the methods, is it the flowing money,[3] is it the intellectual praise? Some of us found disappointment in how similar assumptions about the supremacy of scientific discovery existed in feminist studies spaces. It was not as safe and politically conscious as we had hoped; we gave up monetary benefits and intellectual praise only to find ourselves part of a similar academic elite machine. We found that practicing feminist science in a way that kept our political concerns central was a constant struggle.

What I believe makes many uncomfortable about the metaphor of defection is the reliance on an oppositional, binary relationship between feminism and science. A defector is one who changes sides, reinforcing what many have called out as a problematic characterization of an antagonistic relationship between science and feminism. Instead of such an antagonistic relationship, some imagine our position as between two worlds; for example, Subramaniam uses this metaphor and others of borderland existences (Subramaniam 2005). In analyzing the essays in *Feminist Science Studies* (Mayberry, Subramaniam, and Weasel 2001), a work that is significant for being one of the only readers in this field, I found a common thread of trying to find ways to bridge the two worlds of science and women's studies. Formulating the goal of feminist science studies as bridging two worlds not only leaves intact the binary formation of the two but also tends too much toward a model of seeing the two as equal sides.

Authors in feminist science studies often express a certain kind of feeling of caught between two worlds. Gloria Anzaldúa's metaphor of borderlands ([1987] 2012) may be useful in thinking through what it means to be caught between sides, what the role of borders is, and how the positions of insiders and outsiders are produced and changed over time. Although all positions are complex and multiple, we must carefully examine whether feminists trained as scientists have a claim to finding themselves in a borderland existence between feminism and science. In Gloria Anazaldúa's use of borderlands, power is never absent from the equation. Being located on the border or the wrong side of the border is about understanding the role of borders themselves as places "that distinguish safe from unsafe, to distinguish *us* from *them*. . . . A borderland is a vague and undetermined place created by the emotional residue of an unnatural boundary. It is in a constant state of transition. The prohibited and forbidden are its inhabitants. *Los atravesados* live here: the squint-eyed, the perverse, the queer, the troublesome, the mongrel, the mulato, the half-breed, the half dead; in short, those who cross over, pass over, or go through the confines of the 'normal'" (Anzaldúa [1987] 2012, 25, emphases in original).

Using this definition of borders and borderlands, we might imagine that the feminist might always be in a borderland state in the sciences. However, the scientist moving into women's studies is a different kind of crossing. If we use the U.S.-Mexico border example, which is the literal border that Anzaldúa draws on in her work, an American or Mexican business owner who crosses over to open a new business on the other side of the border is not an equal experience to a Mexican worker who crosses to the United States. My point is that the concept of borderlands is not simply about crossing an unnatural boundary but is about what kinds of normality these boundaries produce materially and discursively. Working from this understanding may allow us to tease apart more specifically the relationship between science and feminism, not assuming they are equal disciplines. When they are treated as equal, I wonder if we are confusing disciplines, political orientations, and epistemological and methodological approaches. This is not to say these do not intersect but that we may fall into slippages between the types of objects we are comparing in our metaphors.

FEMINISM IS TO SCIENCE AS _____ IS TO _____

The interdisciplinary field of women's studies was built by feminists from across the university trained in traditional disciplines in response to student protests.[4] Although there was (and clearly still is) a bias toward humanities and social science disciplines, feminists trained as scientists have also been part of the story of the institutionalization of women's studies in the academy since the beginning (e.g., Ruth Bleier, Bonnie Spanier, and Mariamne Whatley). The questions of how to move from disciplinary knowledge production to interdisciplinary knowledge production and whether and how to interact with original traditional disciplines have been ongoing contentious and challenging issues for women's studies scholars to address.

A variety of answers to these questions have taken shape in the spheres of the humanities and social sciences with some trained in traditional disciplines infusing feminist practices back into their home disciplines by returning to those departments; others have practiced disciplinary knowledge production with a feminist twist within women's studies departments; and still others have worked toward new methodologies that aim for interdisciplinary/undisciplining knowledge production. However, those trained in the Sciences seem to have had a harder time producing similar configurations between their original disciplines and their new feminist departments. This raises a series of questions about why the path for feminist scientists seems to be so different. Is Science itself a special case? Are Science and feminism more incompatible than other disciplinary groupings? And if so, why?

One likely reason for this unique relationship between the Sciences and women's studies is that the methods of contemporary global sciences require expensive equipment and lab space. Due to this, the model for scientific research in the academy relies on what is called "soft money" that is mostly distributed by federal funding agencies. Therefore, using traditional science methods takes a lot of money, requires properly regulated lab spaces, and therefore requires a structural change at a national (and arguably global) level. However, the strained relationship between feminism and science is not simply explained by the financial difficulty of running a proper science lab. If it were, we should find a proliferation of nonlaboratory sciences such as theoretical physics or if we look outside of the sciences proper but within the larger umbrella of STEM, we should find as many feminist mathematics classes as feminist history classes.

So what is holding us back from creating the much sought after feminist sciences? In my introduction to feminist science studies classes, I use an exercise at the end of the course that is adapted from a colleague (Angie Willey). In this exercise, I ask students to define feminism and science separately and then use analogy to explain their relationship to one another. I have them fill in the blanks: feminism is to science, as _____ is to _____. I have found a wide range of responses—some students see them as totally incompatible; others see them as having the potential to be mutually beneficial, others see feminism as a necessary intervention into the sciences, others see them as family members with unequal power, and still others see them as in the middle of a romance narrative ready to come together.

In the feminist science studies literature, we likewise see the use of analogies to explain the relationship between science and feminism. Common metaphors are those of two-way streets and two different worlds. In many of these the focus is on learning to communicate across a language barrier or seeing each other's side. However, these metaphors do not properly illuminate a difference in ideologies between the two or the power differential between science and feminism.

FEMINIST LABS OF OUR OWN IN ACADEMIA? 105

By closely reading feminist scientists' calls for creating two-way streets between feminism and science, I found a conflation of academic disciplines (feminist studies/ science disciplines) with discursive regimes (f/Feminism(s)/s/Science(s)) with Culture/Nature. This is a trend that Angie Willey notes in new materialist arguments (Willey 2016b). She argues that there are slippages between the objects of study and disciplines, so bodies become nature and lead to science being the best disciplinary approach to produce truths about our bodies.

That is not to say there is not a relationship between the academic disciplines, larger discursive regimes, and culture/nature. The easy conflation between these is part of the story of their relationships and how Science has built Nature and gained its epistemic status. Feminist studies as an academic discipline and its relationship to feminism as a counterhegemonic discursive regime are not innocent, unchanging positions either. The struggle to maintain feminism and feminist studies as spaces for counterdiscourse is constant and some would argue impossible (Ferguson 2012; Harney and Moten 2013). I hold out hope and possibility for using feminism and feminist studies as critical spaces for challenging dominant powers. I also do not see this as a pure position. These spaces are not only co-optable but are sometimes predictably and other times unpredictably complicit in producing and maintaining unjust hierarchical systems.

I have always had unease with the way scientifically trained feminists talk about the lack of interest in science in feminism and the difficulty in finding a space for ourselves there—the scientifically trained. This unease probably comes from the fact that the way I perceive my relationship with the field of women's studies is through analogy with the politics of privilege and power. Scientifically trained people have certain unearned, automatic privileges based on their curriculum vitae credentials. You might say, well those are earned privileges then, right? Sure, if we are to ignore feminist science studies critiques of the epistemic authority of science as well as general feminist critiques of the privileging of academic knowledges above others. Scientific merit itself is arguably a way to maintain differences through comparison even as moves to remove bias are implemented (Slaton 2021). If we are discussing our very specific field of specialty, we may have a claim to expertise; however, the idea that we are just generally smarter is problematic. No one comes out and says it exactly like that, but based on experience this is a widespread phenomenon.

The Invisible Pocket Protector

Peggy McIntosh's article "White Privilege: Unpacking the Invisible Knapsack" (2004) has allowed us to unpack a wide variety of privileges and can help us in understanding scientific privileges. I use this analogy to highlight some automatic privileges based on social location (in this case training), but this does not mean that I am analogizing race and academic roles in general.[5] This analogy specifically allows us to think about how epistemic authority traces through to individuals trained in certain disciplines as compared with others.

Instead of a knapsack, I use the metaphor of a pocket protector. This is more of a symbol than an item that is used today by scientists. I choose to use this item rather than something like a laptop bag because of the racialized and gendered specificity of the pocket protector. Laptop bags are used by professionals from business to science to arts, and they are not as associated with a specific gender. I read the pocket protector as a symbol of white nerd masculinity. The physical item, a pocket protector, is generally too large to fit into the shirt pocket of women's clothes, so it is specially designed for male-presenting individuals. If we are to unpack our invisible pocket protectors, we might find . . .

Unpacking the Invisible Pocket Protector

1. I can choose to identify as a scientist rather than a women's studies professor when I meet new people, and they will assume that I am highly intelligent.
2. I can critique science, and it will be taken seriously and not simply dismissed.
3. I can critique science, and there will not be any questions as to whether I actually understand science even if it is a subject matter far from that in which I earned my PhD.
4. I can apply to and get jobs in women's studies departments without ever taking a graduate-level women's studies class.
5. I can teach courses in women's studies outside of science and still be considered competent.
6. I have access to more funding and grant opportunities.
7. I can negotiate a higher salary based on my science credentials (although not always successfully).
8. If I am at a party with colleagues and students, I can act more casually, and my professional credentials will not be challenged.

So, if we see this relationship through the politics of power and privilege, what are our choices as feminist scientists who have some degree of access to both worlds? Could we see our role as supporting the redistribution of epistemic authority instead of being ambassadors between science and feminism?

If we give up on the metaphors of two-way streets and two parallel worlds that *each* need to equally change, we might instead take up the idea of science and feminism as an unequal pairing with science occupying a privileged place. Coming from this position we could see our role as feminists trained as scientists as helping to develop critical science literacy skills based on our firsthand knowledge of the sciences. Perhaps we could create science laboratory practices that are about redefining sciences, both inside and outside the academy. We can be politically strategic about how useful what we are doing is for redistributing power—particularly epistemic privilege through our feminist critiques of science and critical science literacy work.

Before we can do that, we need to be clearer about what we mean by science and what it means to wish to engage with science. While keeping this power differential in mind, I turn to our relationship to doing science through an analysis of our affective attachments and detachments from science.

Our Affective Attachments and Detachments from Science

Over the years, I have often been asked, "Do you miss working in a lab?" When asked this question, I have the sense that the correct answer is that I have that longing—that I have a passion to get my hands dirty in a wet-lab, to work with real matter. How does passion for science, which figured importantly in the do-it-yourself (DIY) labs of the first chapters, show up in feminist communities?

Sara Ahmed's notion of affective economies allows us to think about how passion for science operates through its circulation in feminist studies and in particular through feminist trained scientists. As I discussed in the introduction, Ahmed (2004) argues that emotions are political, not simply individual psychic truths, and that emotions do things. She introduces the idea of affect as sticky in her analysis of how happiness is used in maintaining the assumption of the (heteronormative) family as a social good. She argues that "affect is what sticks, or what sustains or preserves the connection between ideas, values, and objects" (Ahmed 2010, 29). The rehearsal of happiness as the most appropriate affective response to family creates a stronger attachment between the family and social good through what she calls "affective economies" (Ahmed 2004). Consequently, to succeed in this economy of happiness is to be happy about the family and participate in circulating more happiness through love of your family and/or the idea of the Family. Ahmed argues that when one is unhappy about the family, the unhappy person is seen as the problem—the source of unhappiness for all. However, she also posits that we might analyze unhappiness in this case to show that the family is not necessarily a social good. In this way, unhappiness might be able to unstick this relationship. She labels those who disturb these attachments as "affect aliens" or "feminist killjoys." In Figure 4.1, I depict her argument visually showing how happiness is what sticks together the family and social good; unhappiness can potentially break that bond and allow us to see (and challenge) the assumption clearly.

When I began taking classes as part of my ethnographic work in community science labs, I had numerous emotions surface about going back into a science lab. I had not done science since leaving science proper in 2008 after receiving my PhD in neuroscience. I had an overwhelming feeling of anxiety, which I believe was largely based in a fear that I would not be able to do it (science) correctly. The capacity to do and understand science is often thought of as an innate ability, though it is not thought of as an innate ability to carefully move your thumb up and down in the correct sequence for pipetting (see Figure 1.4 for evidence of my pipetting ability). Rather, scientists are assumed to be predisposed to high intelligence, and that intelligence is what gives them the ability to discover scientific truths. In my case, as someone who is often identified in my professional circles as the scientist, I worried about not being able to do science or understand it and having my identity and abilities questioned.

However, I also felt excited about going to the class. I was excited to play with "real" scientific matter, to learn something new, and to try to understand genetics better.[6] I felt excited that working with "real" matter might make possible new knowledges. I had the passion! And in the world of DIY biology, this passion to learn through science is the bodily proof that you belong (see chapters 1–3). I conjured it up, and here I was professing my excitement to tinker and my curiosity about what we could learn from messing with genetic material in a lab space with other geeks and weirdos.

Affect—and specifically passion, desire, and fear—operates to police the bounds of proper science and the questions appropriate to it. A big part of our desire to engage with Science is proving that we (women, feminists) *can* do it. While it is important to counter discrimination by proving that we *can* do science, it is also important to understand and hold space for a *can't* do attitude toward Science.

There are real material differences between the fields of science and feminism. That is, Science is economically advantaged and holds greater epistemic authority. How do we understand our affective relations to science considering this political reality? What does it mean to love, hate, fear, or desire to do science, keeping central the fact there is a material benefit to doing science? How might we strategically position ourselves then in relation to the power of science?

Science is always political and implicated in histories of exploitation. That is, the goal of science is not the innocent goal of understanding nature but has always been about a political project of co-producing nature. I am concerned that we may sometimes forget the political justifications and potential consequences of working toward feminist sciences. Michel Foucault argues that any theory (using Marxism as his example) trying to claim the title of science must carefully analyze what kind of knowledge and whose knowledge it is trying to delegitimize. He argues that we must be aware of the dangers that come with attempting to benefit from "the power-effects that the West has, ever since the Middle Ages, ascribed to a science and reserved for those who speak a scientific discourse" (Foucault 2003, 10). Here it is important that Foucault locates the beginning of this epistemic authority in the beginning of Western colonization.

Sylvia Wynter (2003) draws on Foucault and Aníbal Quijano's conceptions of power to explain that the colonial and capitalist roots and development of the sciences have produced Man thoroughly embedded in a racial hierarchy through what she calls the "Coloniality of Being/Power/Truth/Freedom." We can read Foucault and Wynter together to understand that claiming inclusion into science risks strengthening and relying on a colonial and racialized form of power that is produced by and produces the possibilities for knowledge and life itself. These genealogies demonstrate the co-construction of Science with the globalization of racial hierarchy, which form together naturalized justifications for domination through acts of genocide, slavery, and other forms of exploitation.

At the root of the justification for social inequality, then, is Western Science (together with philosophy and other modern disciplines). By producing categories of human/nonhuman as forms of natural (yet flexible) racial difference, capitalism becomes justified as a natural (yet flexible) economic system (Melamed 2015). We must contextualize our desires to be part of producing science or not within these understandings.

For the *Love* of Science: We *Can* Do It!

In the twenty-first century, we have seen a growth in faculty positions and curricular innovations for feminist science studies in feminist and women's studies departments. As a feminist studies professor trained as a scientist, this has been good news. I noticed something of interest in recent years as I have interviewed, worked, attended conferences, and retrained in feminist studies. My colleagues trained in the humanities and social sciences are not just impressed by my science credentials but often are explicitly interested in hearing the real scientific details of my research.

As I have worked in feminist circles, I have noted that when I am giving talks I am assured that "we can understand the science." I am asked to share more graphs and data so that my colleagues might engage with the science of my work. In some ways this marks an exciting shift in women's studies. It has been much lamented by feminist scientists that our colleagues sometimes act as though there is no way they can understand science or technology (Subramaniam 2005). At the same time, I wonder what is at stake in claiming that we can understand science, and how, in such claims, real science is being defined. Although showing that we *can* do science can point to systematic discrimination that stops marginalized groups from participating equally in science, showing an interest and ability to do science also has to do with the epistemic power that science holds.

Because feminism and feminist science operate within similar cultural spheres as the DIY biology labs that I analyzed in the first chapters, it is not surprising that passion and love for science would be expressed by at least some in these communities as well. By analyzing feminist science studies literature closely over the last decades, I found evidence that love for science and nature were conflated and this affective orientation was used as evidence of a certain kind of naturalness for feminists doing science.

In an exemplary case, feminist biologist Anne Fausto-Sterling (1992) concludes her notable essay "Building Two-Way Streets: The Case for Feminism and Science" by mentioning that she takes great "pleasure from observing the natural world" and is "distressed" to see students not care about the physical world (347).[7] She says that she believes we should be talking about love and pleasure as it relates to science but does not dive into this conversation in her essay. I agree we should be (carefully) discussing our affective relationship to science, nature, and our physical worlds. By beginning her conclusion with this side note about her love of science but not exploring it in detail, this affective claim is left as evidence of a certain kind of unexamined truth about a natural draw to science.

Enthusiasm to do science is an important feature of being capable of doing science in this moment. Therefore, showing passion toward science individually and collectively as feminists is part of showing that we have the aptitude to do science. The shifting landscape of white masculinity has embraced a fun, nerd-brained subject who is passionate about data, numbers, and problem-solving through science while simultaneously remaining socially unaccountable (Willey and Subramaniam

2017). As I argued earlier, passion to do science acts as a colorblind and gender-blind correction to discriminatory assumptions of natural aptitude. However, the idea of a natural predisposition remains as well as an underlying assumption that the intellectual ability to do science is concordant with having the passion to do Science (Willey and Subramaniam 2017). This logic is what not only makes it make sense that some people are naturally predisposed to do science but that others are naturally predisposed to other types of jobs.

Racial capitalism has been the way this is rationalized at group levels and not just for individuals. At the root of it is the idea of innate abilities (or lack of abilities). Therefore, just as we may question the political utility of arguing that some women are as smart as men instead of challenging the idea of intelligence itself, we as feminists may have fallen into the same trap with passion for science. It is common when I am speaking with feminists socialized as women who are not scientists about science that a narrative emerges in which they divulge they had a passion for science at one point, often in their early childhoods, and then that passion was quashed by a teacher, parent, or more diffuse form of power redirecting their energies. These stories, taken together, may be useful in demonstrating that women are still systematically discouraged from entering the sciences, causing the underrepresentation of women in science careers. However, we must not simply show that we too have this thing called passion for science; rather, we must question more deeply what this passion is, how it is produced, how it operates and circulates, and what it means that it is directed toward this object called Science.

Feminist scientists who have defected can play a particular role in telling stories about how and why we left the sciences proper. Even those of us who succeeded to the point of receiving PhDs in the sciences, thereby forever holding paper rights to the title "scientist," but do not continue to practice science are part of the statistics of women who leaked out of the pipeline. The metaphor of the leaky pipeline is commonly used to demonstrate the problem of the underrepresentation of women and certain racial minorities in science in the United States. Many of us begin with stories of passion and love for science and discovery of our natural and physical worlds (e.g., Clarke 2001; Fausto-Sterling 1992; Keller 1977; Weasel 2001). Reading narratives of feminist scientists' complicated love affairs with science may help us to understand more about the political meaning of our affection.

For example, Evelynn Hammonds (1993), now a historian of science who began her PhD training in physics, responds to Aimee Sands in the often-taught interview published as "Never Meant to Survive: A Black Woman's Journey" by saying that she did always "like" science and wanted to do more after being exposed to doing science on her own through a chemistry set that her father bought her when she was a child. However, she also says that she had an interest in doing as much science and math as possible because she knew that would help her get into a "good college." Considering and admitting to the amount of power that science holds in society means that it is impossible to tease out something such as a pure kind of love or passion for scientific discovery outside of the desire for access to power. Hammonds continues to retell the story of how, during what was supposed to be a "fun" summer math program, the two other Black students and she were not having fun because they were not able to keep up due to differences in their previous education compared with the white students. This is an example of how passion and exhibiting a fun desire to do science cannot be considered separate from the ways racism, sexism, and other systems of power operate to differently educate and expose people to what are considered the proper objects and backgrounds in science.

Ruth Hubbard, in response to Fausto-Sterling's call for feminists to be more open to science, points out that the pleasure to observe our natural world is class and race based in our society, where many people grow up without what we consider nature around them (Hubbard et al. 1993). This is an important reminder about the danger of thinking about passion as natural and universal instead of as a nature-cultural process of learning to love (and hate). That is, it is hard to have a passionate attachment (positive or negative) to something that you are not exposed to, something that does not exist in your world. Therefore, if we analyze Hubbard's assertions, we might understand how developing a passion for scientific exploration for nature is itself predetermined based on racism and housing discrimination.

Passion for science renaturalizes who belongs in science while seemingly avoiding the idea of innate intellectual abilities. This reinforces as natural (and fair) the racial and gendered disparities in science. I read Evelynn Hammonds's story as an example of how this process works on an individual level. It is likely that those who can have fun with science already feel/are included, so the requirement to enjoy science is circular in its logic for defining inclusion.

In recalling her scientific beginnings, Evelyn Fox Keller (1977), a trained physicist who ended her career as a science studies scholar, does not invoke a naturalized childhood fascination with science but rather talks about how difficult "origins" are to accurately describe after the fact when she explains how she "fell in love with theoretical physics" in college. "I invoke the romantic image not as a metaphor, but as an authentic, literal description of my experience. I fell in love, simultaneously and inextricably, with my professors, with a discipline of pure, precise, definitive thought, and with what I conceived of as its ambitions. I fell in love with the life of the mind. I also fell in love, I might add, with the image of myself striving and succeeding in an area where women had rarely ventured. It was a heady experience" (78). Keller's story likewise reminds us that it is not just interest and curiosity about our bodies and world that are exciting about science but that we also must understand science as a discipline with epistemic authority. The "headiness" she describes has to do with the power we may gain from accessing this epistemic authority through science, at least in part.

Since Anne Fausto-Sterling published "Building Two-Way Streets" (1992), the metaphor of "two-way streets" is often used to describe the relationship between feminism and science and how to bring them together. As I mentioned earlier, other common ways the relationship is described is as two distinct cultures or languages, and the work of bringing them together is sometimes described as border crossing/working across borders. The common metaphors obscure the power of science as well as the power we may hold as the scientifically trained. A sense of equality and balance between the two is evoked instead. If we consider the analogies to different cultures or worlds more carefully, we will find that there are always power differentials between different cultures. For example, if you are an English speaker, your ability to travel the world without learning another language will be much easier those who speak other languages because of the cultural imperialism of English-speaking countries.

Being strategically clear about our engagement with science might be one way to keep power central while doing science or using scientific results. In an interview with feminist scientist Lynda Birke, who holds a PhD in animal behavior and has since moved to women's studies, Cecilia Åsberg asks whether using proper science is the only means for feminists to get heard and whether Birke sees resorting to this as a problem (Åsberg and Birke 2010). Birke replies, "The trouble is, we simply HAVE no other ways of knowing about the material world that is widely accepted in our culture. That means that we must work with science's authority, even while we might simultaneously challenge it" (418). Here Birke articulates a strategic use of engaging with science. This strategic use counters the language of love and passion for science that naturalizes our attachment to it and naturalizes it as the best way to understand ourselves and our worlds. Instead, a strategic use of it understands that we are talking about science as a specific field or set of methods with epistemic authority in this moment. Birke's answer importantly does not argue for scientific exploration as an ahistorical means to knowledge production.

I suggest here that feminists circulate stories about having childhood passions for Science to disrupt the idea that there are now equal opportunities for boys and girls to pursue science. However, I have shown that a more careful political analysis of our passions reveals a danger in relying on a naturalized idea of passion; instead, I suggest we consider what it means to desire to engage with the epistemic power of science. If not, we risk reifying the idea that Science is the best truth-telling method. That is, we fall into reading all desires to investigate the world around us, our bodies, and our communities as base Scientific passion.

When other options for truth-telling are available, what choices might we make? Would we decide to follow an interdisciplinary feminist methodology instead of trying to be included in the sciences? How might we think about ways that we continued those passions for discovery without science proper? How might we strategically and ethically develop our feminist scientific passions? As I move toward offering some answers to these questions, I suggest we explore the possibility that we *can't* do science.

We *Can't* Do Science: Reading Ignorance, Fear, and Failure as Willful Resistance

If we read our reasons for loving science as more complicated than a pure truth of our innate attraction to the field, what does fear of science or hate or mistrust of science mean, or even scientific illiteracy and ignorance of scientific facts?

In Figure 4.2 Ahmed's idea of sticky affect is applied to understand how Science and Truth are attached through passion and love for Science. The bottom part of the figure visually demonstrates my argument in this section that the absence of that proper affective relationship to Science, through hate, disinterest, and willful refusal, may serve to detach Science from Truth (similarly to how unhappiness may unstick the family from social good).

Fear and ignorance of science are often cited as the primary ways in which women's studies contributes to the impossibility of feminist sciences. When individuals express an antiscience position as a political response, there is often contempt for it. For example, Fausto-Sterling (1992), in discussing the importance of a scientifically literate feminist studies, shares that "at the extreme (but not particularly uncommon) are students who view a refusal of scientific knowledge as an appropriate political stance. To say I find such attitudes distressing is to vastly understate my feelings on the matter" (338). Later in the essay, she continues, "I often experience my students' refusal to explore certain aspects of human knowledge as a willful anti-intellectualism" (339). Fausto-Sterling identifies a political and intentional decision on the part of students to refuse to engage with science. She uses the term "willful" to describe this refusal. She conflates the resistance with a disinterest in knowledge. However, how is knowledge in this case being defined? What does it mean that she is so angered by this political position?

Feminist science studies scholar Banu Subramaniam also uses willful to describe the pattern of refusal of women's studies colleagues to be scientifically literate. She describes a colleague coming to her office to ask about breast milk production as an example. She continues to express frustration at the entire field's ignorance of science as a systematic "willful refusal to be scientifically and technologically literate" (2005, 235). The example of asking an evolutionary biologist who studied plants about breast milk shows us not only a misunderstanding of how science works but a privileging of scientists to know about all kinds of things having to do with science and health, a good example of "reverent disdain" (Willey and Subramaniam 2017). That is, they point to the phenomenon of feminists who have a critique of Science yet continue to believe in its epistemological supremacy, which leads these feminists to count themselves out of being able to do something related to what is considered the large domain of science/nature/technology.

Instead, the willfulness that Fausto-Sterling and Subramaniam describe might be read as an important politically willful resistance to science. Ahmed (2014) argues for the political value of willfulness by producing a willfulness archive in which she shows the ways that this charge has been made against those who threaten the social order, who are disobedient, as a way to dismiss and/or discipline these subjects. Her popular feminist killjoy may be read as a politically willful subject. Although she cautions us that not all acts of self-proclaimed willfulness can be taken as politically righteous, she does give us a reason to pause before denouncing and dismissing an act as willful.

If we go back to my example of the power differences between different cultures, we might think how a non-English speaker living in an English-speaking country

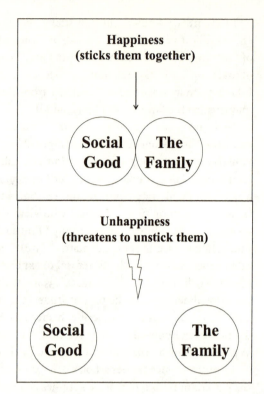

Figure 4.1. Visual depiction of Sara Ahmed's argument about how happiness about the Family acts to stick together the idea of the Family as a Social Good and the opposite, unhappiness, threatens to unstick these ideas.

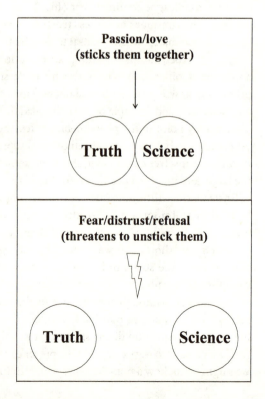

Figure 4.2. Visual depiction of the author's argument that passion and love for Science act to stick together the idea that Science is the best arbiter of Truth and how negative associations such as fear, distrust, and refusal toward Science may unstick these ideas.

FEMINIST LABS OF OUR OWN IN ACADEMIA? 115

and refusing to speak or learn English may be read as a political act of willful refusal against cultural imperialism. To be illiterate in this case might be an important act of resistance. Moving this analogy back to thinking about science illiteracy, we may find that the actors are intentionally providing a different language or way of understanding their bodies or worlds in contrast to the dominant forms of science. In Jean Barr and Lynda Birke's (1998) work based on interviews with women about adult science education, they analyze what is commonly seen as ignorance of science as potential resistance. Barr and Birke found that sometimes the women they interviewed knew the scientific answers but chose other nonscientific explanations to describe the natural world. They also point out that Emily Martin (2001) similarly found that what might at first appear as scientific illiteracy among working-class women was actually a resistance to medical authority on the subject of menstruation. Therefore, Barr and Birke argue, women are not just "alienated" but there is also an "active resistance" to science proper.[8]

Perhaps these examples could be thought of as a kind of biodefection. Ruha Benjamin (2016) introduced the idea of informed refusal as a counter to informed consent in the medical sphere to honor the kind of work that those resisting medical authority do for remaking knowledge and being through what she calls "biodefection"—a refusal to rely on biological claims to understand ourselves and our worlds. This concept has power to renarrate what is commonly called noncompliance within medical spheres. I understand noncompliance with medical authority as related to the disdain toward those exhibiting science illiteracy. There are many reasons why we might wish to politically resist the dominance of science (and medicine) as well as ways that we are forced outside of traditional sciences.

Both Ruth Hubbard, a trained biologist and notably first woman to be tenured in biology at Harvard University, and feminist philosopher Sandra Harding respond to Fausto-Sterling's two-way streets essay by taking issue with the way Fausto-Sterling talks about women's chosen ignorance about science (Hubbard et al. 1993). Hubbard says that she is also concerned that we (feminists) do not know enough about science, but she does not think this is about a choice as much as a result of the exclusion that has happened and that the language itself is difficult to understand. Therefore, she points out the systematic ways that we are kept ignorant about science instead of placing the blame on feminists. Harding addresses this issue of science illiteracy in her response to Fausto-Sterling's essay by zooming out and saying we need to look historically at the ways that this knowledge has been systematically kept from some groups: we need to shift most of our focus away from how groups who are illiterate need to get over history and get interested and onto what is so wrong with Western sciences.

It is vitally important to remember that many of us were unable to do science because it is a space that we were denied full access to. Bonnie Spanier, an emeritus women's studies professor trained in microbiology and molecular genetics, offers an example of this in her retrospective of the impact of a decade of women's studies on the natural sciences:

Among the voices of women's liberation heard in the late 1960's and very early 1970's were only a few feminist scientists. Gradually, personal accounts understood through feminist consciousness appeared, breaking through a strongly imposed silence about the real conditions of many women's experiences studying or doing science. Early feminist writings illuminated the dark side of women's experiences in science in direct ways that were avoided in arenas closer to the scientific community. In 1970, Diane Narek gave her view of the problem of women in science:

"The only reason there aren't more women scientists and technicians is because the men won't allow it. They tell us women that we are not good at mechanical skills. If we disprove their theory by learning these skills, they accuse us of being unfeminine. As a youngster I was encouraged in my scientific pursuits, yet when I tried to achieve these goals I was harassed. *I had the confidence in my ability and the desire to study science.* The constant pressures that I received caused me to lose both. . . . *It was only after my involvement in Women's Liberation that I again found enjoyment in science.*" (Spanier 1986, 67, emphases mine)

Spanier further explains how difficult it was for women in science to speak out for fear of being delegitimated within a field where they already struggled for legitimacy. Here, Spanier's retelling is important to remember. Not only were women scientists routinely discriminated against in the sciences, but they were also made to feel shame about their failure. It took feminist political community for women to understand this as part of a larger problem and not a personal failure. Failure could be not doing experiments properly, being criticized when presenting work, not gaining employment or receiving grants, not publishing enough, and being thought of as not smart enough to participate in science. While these failures are posed as individual shortcomings based on an individual's aptitude and/or effort, the pattern of women failing at these more often than men points to ways that these measures are systematically set up to disadvantage women.

A striking anecdotal example of how a woman's research versus a man's research is judged is Ben Barres's experience as both a woman and a man in science. Barres is a transgender scientist who shared his experiences in *Nature* to help dispel notions of natural differences in aptitude between males and females (Barres 2006). He recounts a presentation in which he overheard other scientists afterward saying that his work was much better than that of his sister. The assumed sister was the same person, Ben Barres, before he transitioned to male. Studies continue to be conducted and published showing that employment, admissions, and publication outcomes change based on whether the person's name appears to be male or female and based on racial assumptions.

Feminism has provided a way for those isolated in the sciences and made to feel personally responsible for their failures to instead understand the systemic pattern of discrimination. For example, Keller (1977) discusses her "own rather painful and chaotic history as a woman in science" (78) as a refusal to stay silent and internalize it as a personal failure. I read Keller's story as an example of her "affective dissonance" turning into politicization. Clare Hemmings (2012) argues that an affective dissonance between what you believe you can do and what is politically and socially possible can be (although is not automatically) a precondition to a politicization that seeks to change that gap. Keller's involvement with the "women's movement" seems to be a part of what gives her the power to take her feelings of "shame" along with "anger, confusion, and timidity" (78) that had stopped her from speaking out previously and move to a place where "it seemed decidedly inappropriate, somewhat dishonest, and perhaps even politically unconscionable" to not speak out about what she had experienced as a woman in physics (78). She describes the lack of support and investment in her after she had been admitted to graduate school, the isolation from her peers, and the comments from professors and students about her inability to be as good as she thought she could be at physics. She then describes her politicization through an affective change: "My shame began to dissolve, to be replaced by a sense of personal rage and, finally, a transformation of that rage into something less personal—something akin to a political conscience" (80).

Ironically, finding a feminist home to process the systemic lack of fairness in the sciences has provided the impetus for several of us to actually continue to work in the sciences. The story related by Narek (1970), as reprinted by Spanier (1986), is not alone in crediting feminism with giving her the ability to continue to do science after affective dissonance in the sciences proper. Banu Subramaniam (2005), who was trained as a plant evolutionary biologist, and Deboleena Roy (2004), who was trained in reproductive neuroendocrinology and molecular biology, likewise give credit to women's studies and feminists, respectively. As I mentioned at the beginning of the chapter, this was my experience as well—that is, it was members of a feminist reproductive justice organization who implored me to continue to finish my degree and to ultimately aim to conduct science for social justice.

Our collective stories also show, however, that once we do find a way to reconcile our politicization with "doing science" it is not simple or necessarily possible to do science the way we want to within the proper spaces of science. Many have found that once you start to analyze Science itself and ask questions, you are not welcome and those questions are not welcome (see Hubbard's comments in Hubbard et al. 1993). This was true of my own experience as I described at the beginning of this chapter in searching for a postdoctoral fellowship after finishing my PhD in neuroscience (Giordano 2014).

Another example of this comes from Roy (2004), whose primary appointment is now in women's studies but holds a joint appointment within the biological sciences. In her own attempt to incorporate feminist theory and practice into her doctoral work, she found that there were compromises that had to be made along the way; that is, it was not possible to complete the kind of feminist project she would have liked to because of the structure of the scientific field. Certain questions and topics were considered not appropriate for her thesis. She recounts, "As a graduate student in the sciences, I quickly learned that 'radical' and 'nonscientific' ideas would not be tolerated and that intellectual conformity seemed to be the key to success."

This experience is echoed by Caitlyn Allen (2001), who received her PhD in plant pathology and also still holds a joint appointment in natural science and women's studies, as she identifies the difficulty of being critical of science in science: "An interest in the philosophy of science or a critique of scientific approaches is widely viewed as weakness (except in emeritus faculty, where it can be a charming eccentricity). The taint of dilettantism can be fatal for a young scientist; one of the most damning things one can say about a student is that she's not serious about science" (24).

These examples point to a problem. We *can't* do science the way we need to once we begin to be exposed to feminist critiques of science. There are systematic ways that we are stopped from pursuing the kind of feminist science questions that we want to and using feminist methodologies, such as working directly with the most impacted communities in developing projects (Giordano 2014) or considering relationships between research animals, subjects, or matter and the scientist (Roy 2004).

What kind of science can we practice, then, once we are pushed out of the proper domain of Science? There are roles for those of us trained as scientists in education and redistributing epistemic authority and perhaps even resources allocated for scientific exploration.

INTERLUDE 5

When the Right Comes to the Defense of Science

Parts of chapters 4 and 5 appeared in an article in the journal *Catalyst: Feminism, Theory, Technoscience* in the fall of 2017 (Giordano 2017b). I was surprised when I received an email shortly after the article was published from a reporter asking for an interview. Someone had read my article! I looked up the journalist's publication and quickly found out that it was the (now-infamous to many in the academy) right-wing news site *Campus Reform*. I did not respond to the request, resulting in a final sentence of the article stating, "Campus Reform reached out to Giordano multiple times for comment but did not receive a response in time for publication."[1]

After the critique was published, I began receiving angry and nasty emails about my article. Most of them were quite short, just stating that I was stupid and/or harmful. Although many targeted by *Campus Reform* have received threats of physical harm, including death threats (Tiede et al. 2021), most of my own anti-fans stuck to insults. There was a subtle threat of sexual violence in one email that left me wondering whether the letter writer had researched me beyond the original post enough to assess my queerness (the suggestion was that I needed sex with "a man" to correct my thoughts/way of being).

The most fearful moment was when I received physical mail delivered to my office, and inside the envelope was a Chick tract with my name written on it. Chick tracts are small evangelical Christian comics with anti-Islam, anti-Jewish, and/or anti-Catholic messages (Wilkinson 2016). The one I received was aimed at convincing the reader of the devilish ways of Muslims in an apparent effort to convert me to the "good" side of Christianity. Knowing that someone did not just click on an email address but went far enough to look up my physical address caused more fear of safety than the previous messages. I also was placed on a watchlist of "liberal" and dangerous professors, and multiple other right-wing online publications republished versions of the *Campus Reform* article.

I am reminded of the emotional distress caused by this incident as I write this. I have a deep unease that makes sense: the battle over knowledge and truth has

119

never been limited to text. This is what makes our work so critical and dangerous in the full sense of material-discursive ways. Ironically, this attention also brought me into a community with a larger network of others who also have been targeted; they have used these lists and right-wing publications to find allies in the fight for social justice.

The reason I include this introduction is to analyze how an article that was aimed at provoking conversation within my very small community of feminist science studies scholars gave me more exposure outside the academy than any other work I have put out so far. Perhaps this book will make it into a future issue of *Campus Reform* as well!

Readers of both my original *Catalyst* article and the drafts of these chapters have shared concerns that my arguments could be used by the right wing and would play into the hands of anti-Science thought, which has been linked with increasingly racist politics and the denial of climate change. After all, I had planned to publish this work less than a year after Donald Trump had taken over the U.S. presidency after a populist campaign that had drawn heavily on the intersection of racism, Islamophobia, and climate change denial.

Some have suggested that the rise of postmodern thought and challenges to the idea of "truth" have allowed for the rise of dangerous anti-Science thought. I posit that this is a simplistic, misguided explanation. Likely the rise of academic challenges to Truth were and are entangled with larger social/cultural challenges to Truth—meaning that the challenges are not unidirectional. One does not simply cause the other. In addition, as I noted in the introduction to *Labs of Our Own*, considering pro-Science and anti-Science to be opposite, binary logics is a fallacy. Overreliance on the idea of these as two discrete positions with corresponding good/bad politics mapping further onto antiracist/racist or environmentalist/antienvironmentalism is a mistake.

The enemy of your enemy is not necessarily your friend. If some use a critique of absolute truths passed down by an elite academy to further a populist, sexist, racist agenda, this does not mean that we must embrace ideas of Academic Truth. By publishing my critique of our affective attachment to scientific objectivity as firmly grounded in its co-production with Western colonization and racial capitalism, right-wing activists quickly came to capital-S Science's defense. Hence, supposed left/right alignments with Truth and Science are not as simple as we may think/act.

Perhaps instead of trusting Science more, we need to trust all our abilities to process complexity and more democratic knowledge production processes. With this in mind, I next suggest some concrete steps toward feminist science, but not before I more fully consider the potential of failing at science—and actually listening to the arguments against Science—before dismissing them as "antiscience."

CHAPTER 5

Toward Queer Sciences of Failure

In *The Queer Art of Failure* (2011), Jack Halberstam argues, "Under certain circumstances failing, losing, forgetting, unmaking, undoing, unbecoming, not knowing may in fact offer more creative, more cooperative, more surprising ways of being in the world" (2–3). Failure has taken up an important space not only in queer studies but in science as well. Many have called attention to the ways experiments that fail to prove the hypothesis correct disappear from the scientific record. There is at least one journal dedicated to publishing nonsignificant results, the *Journal of Negative Results*, drawing attention to this problem.

Some scientists argue that reporting on negative results is important because these rule out certain possibilities and can save others time when they do not need to retry something already tested. Also, with reported data skewed toward successful studies, it may not be clear when results were attained under particular circumstances and are not as reproducible as thought. If only studies that show positive links between assigned sex and behavior are reported, we miss out on the numerous times that those factors are attempted with no significant correlation was found. This leads to a bias in reporting and ultimately in our beliefs about the significance of sex differences. Of course, simply reporting or embracing failure will not in itself correct unjust scientific practices or findings. An embrace of failure through the trial-and-error practices of the tinkerer has been developed through a neoliberal framework, thereby reproducing similar social hierarchies and ways of valuing scientific knowledge production.

An alternative form of failure would pose an explicit challenge to capitalism. Halberstam (2011) argues that we can read a history of successes and failures during capitalism in multiple ways and suggests reading the history of failures as "a tale of anticapitalist, queer struggle" (88). The word "queer" suggests a challenge both to capitalism and to the categories of human/not human and normal/abnormal that Science has enforced and drawn on for success. A new anticapitalist science in and of itself will not necessarily "unsettle" (Wynter 2003) colonial, racial, and gendered power. Wynter's formulation of not simply the Coloniality of Power but the merging

121

of Being/Power/Truth/Freedom clarifies that what is at stake in unsettling coloniality is the definition of the human itself.

Queer theory can be particularly important in science education because of its concern with the body, critiques of biological explanations, and opening up of new possibilities for being (Britzman 1995). However, this point is also an important place of shared concern between feminist science studies, critical ethnic and Black studies, and decolonial and postcolonial studies (Pollock and Subramaniam 2016). The question for those of us in feminist science studies is how to make sure we take seriously the insights and practices of scholars in these critical disciplines; if we do not, we risk reifying racist and heteronormative approaches that leave intact the deep inequities built and maintained through colonial projects.

In this vein, we may heed Wynter's (2003) call for a new science to produce a new "descriptive statement of the human." Intimations of this "new science" appears in Wynter's earlier work (Wynter 1984) in her call for "rewriting knowledge."[1] This new science/rewriting of knowledge is about undoing disciplinary boundaries and Truth that produces "narratively condemned status[es]" of Others/nonhuman humans (Wynter 1994b, 70) and "unsettling" the relationships between Being, Power, Truth, and Freedom stuck together through processes of colonization (Wynter 2003).

What can we gain from reading feminist failures in science differently? The metaphor of the pipeline is commonly used to discuss the attrition of women and underrepresented minorities from the sciences. Banu Subramaniam (2009) has critiqued the use of this metaphor by asking whether the metaphor assumes that our goal should be to keep people in the pipe, ignoring the fact that the pipe itself may be the problem. She asks whether a reorientation to the metaphor might have us celebrating those who escape the pipe instead. This is a recurring concern in feminist science studies literature. Lynda Birke (1991) also critiques trying to make people fit better into a science that is not meeting our needs through science education: "Many people perceive things to be not quite right in the science camp, while scientists and educationalists are wondering how to persuade more people to pitch tents there." The question for feminist science studies scholars then is, why are similar concerns being brought up twenty years apart? Why do we find ourselves once again being concerned with science proper and often striving for working with "real" scientists? Why are we still unable to influence diversity initiatives aimed at increasing women and racial minorities in the sciences?

Reading critiques of simple inclusion politics together with Halberstam's (2011) *Queer Art of Failure* raises related questions: have feminist scientist defectors left the science camp, broken out of the pipeline, and produced a queer science of failure that feminist science studies scholars should celebrate? And if so, then how should we read desires and calls for laboratories of our own so that we do not simply continue building that same pipeline through new territories, colonizing more area by pitching science tents? By focusing on the political consequences of our arguments, we can use them to say we need laboratories of our own to fulfill

In Defense of Science Illiteracy and Unknowing Science

our passions to produce knowledges about our worlds and bodies in such a way as to broaden science rather than simply gain inclusion in Science as is.

In Defense of Science Illiteracy and Unknowing Science

Many scientists and policymakers focus on how to get more girls, women, and underrepresented racial groups to participate in science—particularly when they want these groups in clinical trials. The people who are refusing to participate in clinical trials or treatment are often seen as lacking sufficient trust in institutions. Simply blaming these groups for the lack of participation or even sometimes accounting for specific abusive histories leaves the larger history of science unquestioned and ignores continued exclusions and abuses (Washington 2006). Instead, the distrust might be warranted, and the appropriate response may be to interrogate the trustworthiness of the institutions themselves (Benjamin 2014, 2016). We should examine how the separation of those who control the means of knowledge production (researchers, doctors) and those who are the subjects of knowledge production (research subjects, patients) maintains what Harriet Washington (2006) calls a "medical apartheid" system based on race. Black people across genders remain severely underrepresented in the sciences, but white women (often benefiting from affirmative action programs) have become a large part of the sciences proper, specifically in biological sciences.

In critical studies there have been more and more analyses that aim to understand politics of refusal and resistance. In the sphere of medicine and science, particularly in this political moment perhaps when science and medicine are being positioned as enemies of a racist, Christian right agenda, some may have mistakenly seen the enemy of their enemy as their friend. In this formulation there are two sides: proscience or antiscience. As feminists and critical scholars, we know that assuming there are two sides is a problem. However, we need more language to read the kinds of antiscience stances taken by activists who are antiracist, feminist, and queer. These forms of resistance are often well-thought-out approaches that do not fit a simplified antiscience position. The use of the term "biodefection" by Ruha Benjamin (2016) to describe those who have been traditionally harmed by medicine opting out of a biological citizenship that requires participation in genomic medicine is one example of a much needed analysis of medical resistance.

We can also turn to those at the borders of activism and academia. Independent artist, film maker, and activist Lucía Egaña Rojas (2013) writes in "Notes on a Transfeminist Technology" that "a transfeminist technology will value illiteracy for its improductiveness for industry, as a way of finding paths unimagined by speed and productivity" (27). Rojas further suggests a usefulness in creating new worlds by being gender illiterate as well as an acknowledgment of how the positive relationship between epistemic power/authority and literacy devalues the knowledge of many of the world's poorest inhabitants. Rojas's exploration of illiteracy opens a space for the importance of science illiteracy as a failure that threatens the supremacy of Scientific knowledges.

What kinds of science illiteracy might we embrace as part of a project of destabilizing Science and remaking knowledge production? When decolonial activists in South Africa who were part of the Fees Must Fall movement (which stemmed from Rhodes Must Fall) suggested a total abolition of Science before being able to remake new sciences, they came under intense fire. A video of one activist speaking at a university event about science went viral on social media in October 2016 with the hashtag #ScienceMustFall.[2] The commentaries online showed that the claim of attacking Science was used to delegitimize the activists' movement. The suggestion that the activist made about Science being a colonial project and needing to be abandoned before creating something new was not just rejected but taken as evidence of ignorance. The activist also makes mention in her brief comments that she was once a science student and suggests that it is because of the boundaries of what counts as real science that she did not continue in the sciences. She is another one of the science failures we must listen to and engage with as we rethink a feminist engagement with S/science(s), one that takes an antiscience approach to creating new knowledges while reclaiming science teaching and activist sciences as science.

These ideas are not theoretical exercises. As I write this, we are in the middle of a pandemic that has claimed the lives of millions of people around the world. We are dealing with arguments over truth and science daily, and these conversations are not limited to the academy or simply activist circles. For example, truths about vaccination as a solution to the pandemic have become politically volatile territory. If we had a more nuanced discussion about distrust of Science, perhaps the debates in 2021 about COVID-19 vaccines would have been less of a shock to well-meaning liberals in the academy and beyond. Maybe feminist science studies scholars and activists could be useful in this moment, not in pushing for greater "trust" in the Science and the biomedical industry but as being trained and practiced in understanding evidence in a contextualized way. Perhaps we could not just be useful in convincing "the uneducated" to trust the vaccine, but maybe we would all learn from a process of knowledge production rooted in democratic, liberatory education.

THOSE WHO CAN, DO; THOSE WHO CAN'T, TEACH

Many of us defectors have not been able to return to lab bench science. We have continued to engage in science through our teaching. We can read the popular aphorism meant to devalue teachers—those who can, do; those who can't, teach—through Sara Ahmed's (2014) work on willing and willful subjects to give it a different power. The failure to do science might be seen as a willful political gesture against a science that reproduces injustices. The defectors among us could not *do* it any longer. We often tried to stay in the sciences and make change from within, but at some point we realized the limits of our work from within was as feminist spies.[3] When being a feminist spy was not enough, we had to imagine other ways of becoming knowledge producers. We turned to teaching critical science studies, thus creating our own feminist science labs in feminist studies classrooms. And by turning to teaching, specifically teaching in feminist studies, we not only fail at being a part of Science proper but are failing capitalism.

In *The Queer Art of Failure*, Halberstam uses art "to think about ways of being and knowing that stand outside of conventional understandings of success" arguing "that success in a heteronormative, capitalist society equates too easily to specific forms of reproductive maturity combined with wealth accumulation" (2011, 2). We defect and choose to take lower-paying jobs, devalued not only in economic terms but in social status and power. Perhaps claiming that we choose this route of failure (teaching) is giving too much agency to us without acknowledging the historic ways that people of color and women are overrepresented in these underpaid professions. As I point out in the previous chapter, many of us cannot do Science because we are systematically kept out of the institution. By pausing to embrace this queer failure, and through this kind of teaching and delegitimizing of Science, we open possibilities to create new definitions of science—ones that are politically grounded in social justice and the politics of location—so we are therefore also doing or creating new *sciences*.

Subramaniam asks, "What would it mean for women's studies to engage with the sciences as its own? Not at arm's length, not with fear, not with paranoia, but owning it as ours to shape, to empower? What would it mean for us to have laboratories of our own?" (2005, 238). To answer these questions we need to acknowledge that the reasons for fear of science are real and not paranoid delusions of potential harm. I like the idea of owning science as our own. But who is "our" in this case? Is it those of us with a privileged scientific training, which also means a predominately racially privileged group? How might we use this privileged position, holding onto why we *should* be fearful of Science and remaining politically aware of our own potential reifying of unjust power in utilizing science? How might we then move forward with feminist sciences?

Evelyn Fox Keller (1977) describes the transition from a traditional research position to teaching as providing her with the ability to pursue knowledge production in a way that was satisfying to her political and personal commitments. It is not only on an individual level that teaching science from a feminist perspective might be fulfilling; teaching is a political project of great importance and a significant focus in the field of feminist science studies for that reason. The *Feminist Science Studies: A New Generation* reader is broken into four sections with one full section about "pedagogical and curricular transformations" (Mayberry, Subramaniam, and Weasel 2001). In this reader, multiple science-trained feminists discuss their teaching even outside the section on pedagogy (e.g., Allen 2001; Weasel 2001). Nearly all feminist-trained scientists have published on education or pedagogy in some form (e.g., Barad, Birke, Giordano, Hubbard, Roy, Weasel, and Whatley), a testament to the large focus on pedagogy and education in the interdiscipline of feminist studies. What does this attention to teaching and pedagogies mean for producing new scientific knowledges?

Laboratories as Classrooms/Classrooms as Laboratories

Feminist scientists who have defected from traditional scientific lab spaces offer a special kind of insight to the world of science. They have studied it from the "inside" and now "outside." Defectors can aid the other side with firsthand intel on what their former allies are plotting, how they do what they do, and where their weaknesses are. Uncomfortable as this militaristic metaphor might be, it highlights that feminists who have been trained as scientists do indeed hold claim to an epistemic advantage in understanding the ins and outs of the sciences—perhaps more so than current scientists who have not challenged any of science's assumptions or nonscientists who have not experienced it firsthand. To this end, Deboleena Roy (2004, 2012) and others (including myself—Giordano 2014, 2017a) have attempted to create practical advice to feminists and in particular feminist scientists in how to engage with scientific knowledge production though our writing and teaching.

Often, we have been able to bring science to feminist studies through our pedagogy. I call this work feminist critical science literacy. Others have also taught students across disciplines to become responsible knowledge producers (Barad 2001) and how to read and engage with primary scientific literature (Roy 2012) as a feminist method. Many seem to see this as a step toward doing "real" science, toward creating "labs" of our own instead of a laboratory in and of itself. Yet by viewing these practices as producing new kinds of scientific knowledge in and of themselves, we can widen the definitions of laboratories and science.

Lab, as I noted in the introduction, has many different meanings and functions beyond simply a laboratory space. In this case, laboratories are not simply spaces but also a pedagogical model. Laboratories in the sciences are (1) classrooms for learning methods and doing predetermined experiments to learn or demonstrate scientific principle and (2) places of research (and labor) to find new truths. Students in science classes from high school to college are most familiar with the first meaning of laboratory. Pedagogically, there is an assumption that the hands-on experience in the laboratory is important for knowledge acquisition. What else do we learn through practicing "being" in laboratories? What kind of affective attachments or detachments from science does the laboratory experience produce? The firsthand doing of science to learn scientific principles reinforces the materiality of really doing science and perhaps the materiality of science itself. If we understand feminist science classrooms as laboratories, what could that do for our ideas of science, scientists, materiality, and truth?

The classroom laboratories of high school and college are also places of disciplining knowledge-seeking practices. Feminist classroom laboratories instead can be seen as places of potential openings of new knowledge seeking practices. In traditional classroom laboratories, the repetition of the same experiments and the practice of learning to "write up" labs in your lab notebook are disciplining students' bodies to "do" science. In feminist classrooms, we instead work toward undisciplining science as part of a resistance to traditional scientific truths. This is the work of affectively producing an irreverent disdain for science.

The notion that our teaching *is* about producing new scientific knowledges is not necessarily a new claim but one that has perhaps lost prominence and is sidelined by comments about moving beyond critique in feminist science studies. A multisite National Science Foundation (NSF) funded project that took up Fausto-Sterling's two-way streets metaphor as a project title focused primarily on pedagogical interventions and provided specific examples in a lengthy follow-up report to the project (Musil 2001). In the introduction to the report, Caryn McTighe Musil (2001) explains that this differs from typical science literacy projects because these programs work on changing science itself.

Many of us scientifically and not scientifically trained feminist studies professors may be teaching science and engaging with science without realizing it. For example, as part of the NSF funded two-way streets project, Bates College faculty found that the reverent disdain that women's studies faculty exhibited toward science at first meant that they saw science as something that needed to come from outside their classes and research (Baker, Shulman, and Tobin 2001). However, they found, to the surprise of organizers and faculty participants, that they were already engaging with science (broadly and critically defined) in their teaching and research. For example, most feminist studies professors will teach about reproductive justice and social constructions of sex/gender, disability, and/or race. Teaching these topics requires that they challenge scientific truths about difference and provide new ways to understand our bodies and communities.

In Maralee Mayberry and Margaret Rees's (1999) reflections on a co-taught course, they define pedagogy as *doing*, using the definition of Frances Maher and Mary Kay Tetreault (1994) of pedagogy as "the entire process of creating knowledge, involving the innumerable ways in which students, teachers, and academic disciplines interact and redefine each other in the classroom, the educational institution, and the larger society" (Mayberry and Rees, 1999, 196). They argue that feminist pedagogy "at its core . . . [is] a commitment not only to interdisciplinary knowledge and process of learning but also to the development of a critical consciousness empowered to apply knowledge to social action and social transformation" (206). This sets a basis for seeing teaching as producing new kinds of knowledge that may change our very definitions of the human and accordingly our understanding of the worlds we inhabit. These definitions change power and change the material conditions of our lives.

Critical science literacy is not simply learning how to read or engage with science but is doing science itself. The most foundational science studies tenet is that all scientific knowledge is socially situated. That is, science is not *ever* simply produced in scientific laboratories.[4] Feminist philosopher Helen Longino (1990) uses this principle to argue that we cannot have feminist science until we change culture itself to be feminist. Literacy also needs to be understood as not simply the ability to read but also, as defined by language specialists, the ability to write, listen, and speak. Although basic science literacy has not been taught as a way for the public to engage in science-making but rather in science-believing (often through appreciation), feminist critical science literacy is part of a tradition of producing responsible scientific actors across disciplines (Giordano 2017a). Why do we see science literacy as simply reading or understanding science rather than the ability to write it and remake it? In the vein of Wynter's call to rewrite knowledge, critical science literacy offers a way to rewrite scientific knowledge.

The idea of broadening the responsible actors of science is popular among feminist science studies scholars. Feminist science studies scholar and trained physicist Karen Barad's essay on science literacy, "Scientific Literacy → Agential Literacy = (Learning and Doing) Science Responsibly" (2001), argues for extrapolating her concept of agential realism to scientific literacy to learn to intra-act responsibly within our worlds. She explains that the course she developed "was designed to enable students to learn science while thinking about science, and to learn that thinking about science is part of doing science" (240).

Similarly, Jean Barr and Lynda Birke in *Common Science?* (1998) define a critical science education that has to do with broadening our definition of those responsible and who should take part in science: "A critical science education would, in consequence, involve working with women's groups in the community, drawing on their own agendas, whether to do with housing, health, roads or the environment, in an effort to develop more broadly based 'scientific communities.' The kind of science education we envisage here is an aspect of citizenship education of 'pedagogy through politics' rather than a pedagogy centered solely on the classroom (see Le Doeuff 1991)" (136).

However, claiming science literacy as science on its own is insufficient. We must carefully define our reasons for the need for science literacy. Specifically, science literacy can be used to produce knowledge about our worlds and bodies through a redistribution of epistemic authority. There are good reasons to resist the authority of Science. However, there are also good reasons for science literacy and engaging with science so that we are not leaving "the terrain of science uncontested" (Barr and Birke 1994, 482). An example of the danger we face in not addressing science comes from an early feminist studies scholar Mariamne Whatley (1986), who was trained in the biological sciences. She argues that if we do not deal with biological determinism at all in our classes then we "create more vulnerability to these theories and . . . give more power to those who use these arguments. These issues must be dealt with openly in all women's studies classes or the biological arguments will continue to carry too much weight" (186). It is easy for us to slip from this position into a position of critiquing and blaming those who do not engage with science as the problem. A more strategic and ethical position would be to hold both resistance through nonengagement together with engagement with science through rewriting scientific knowledges.

We can challenge the epistemic authority of Science by not only learning critiques of its larger philosophical underpinnings but also learning to read primary science articles carefully to critique methods and results of specific studies. In this way, we can participate in the day-to-day practice of science. However, there are physical barriers to finding science articles as well as the difficulty in the specialty of the language that is used in writing about science. One practical and useful role for science studies scholars, especially those trained as scientists, is to make it possible for more people to understand scientific practice and writing. Once again, this is not a new idea. Ruth Bleier (1986), a trained medical doctor and neurophysiologist, argued that "it is also our task to make science, and the feminist and other radical perspectives on science, accessible to new generations of students within women's studies and science classrooms" (2). Whatley (1986) argues, "What is more crucial than just supplying more accurate information to use as ammunition in debates is to help students develop alternative hypotheses, to see the roles social, cultural, and political factors can play in what appear at first to be biological issues" (186–187). She gives examples of how this can be important if incorporated in physical education classes to help students develop an understanding of physical ability, hormones, strength, and the social aspects that co-produce each of these concepts.

More recently, Deboleena Roy's detailed methodology for reading and engaging with science was included in a handbook for feminist methodologies. She argues, "Feminist scientists will recognize that learning how to read a scientific paper *closely* is indeed a much needed feminist practice in the natural sciences. It is extremely important for the feminist scientist to learn how to read peer-reviewed scientific articles as texts and look for the possible ruins and risks entangled within these texts" (2012, 25). She continues with a close reading and reinterpretation of a study on gender and humor, demonstrating critical science literacy by engaging and producing new truths about this subject. Although she often speaks directly to feminist scientists working at the lab bench, her methodology is significant for feminist scientists working in women's studies laboratories/classrooms as well.

REDEFINING THE FEMINIST SCIENTIST: WHO IS *OUR* IN FEMINIST LABS OF OUR OWN?

A certain amount of failure at following the rules (for some people) can be seen as brave and generative in the sciences. For example, the image of the rogue scientist (who is generally a white male) remains an integral part of capital-S Science knowledge production. But simply being outside the academy or having failed at disciplinary sciences cannot be our criteria for queer sciences of failure. Darwin and many wealthy gentlemen were the first scientists in this tradition where they used their power not to study what was in the best interest of most people but to follow their independent (bourgeois) scientific questions.

This image even appears in Halberstam's introduction to *The Queer Art of Failure* (2011) where he uses the rogue (lone) scientist/mathematician who works outside of academia as an example of the possibilities opened up by undisciplining knowledge. Although his book overall complicates the meaning of failure rather than simply celebrating it, the introduction makes a strong case for how queers, feminists, and revolutionaries have and might want to embrace failure over success in a capitalist system.

I wondered why the lone scientist was seen as compatible with these political factions. The lone scientist can be understood as part of a long-standing colonial narrative that begins with the navigator Columbus whose error/failure has been celebrated. Today, the lone scientist describes several synthetic biologists and more broadly the biotechnology stars who are celebrated for their entrepreneurial risk-taking outside the academy. This is a reminder that failure at traditional sciences must always be politically contextualized.

As I discussed in the last chapter, feminists trained as scientists benefit from the epistemic authority that Science holds. Feminist scientists must therefore take a lesson from privilege and power politics and think about how we can redistribute epistemic authority through our privileged position as the scientist. At the same time, we need to work to dismantle the privileged status of the trained scientist— that is, question who counts as actually doing science and who is thought to be capable of producing scientific knowledge.

Lab assistants and students often do the day-to-day labor of scientific research. Laboratories are not simply places for elite knowledge production and teaching but places of laboring.[5] When it comes time to take credit for scientific discoveries, those with the most power receive the credit. Even in science tinkering communities that are based in the idea of hands-on science experimentation, much of the labor is outsourced to unnamed laborers. What could we learn from those who do the invisibilized labor of science? Roy (2004) drawing on Spanier's methodology offers that part of a feminist methodology would include considering the relationship between the scientist and the material. We might, in fact, consider all the stages of labor that bring the material to the lab bench scientist. For example, we should question whose labor produced the equipment that we are using, from where the materials were extracted, what might be the extraction impact on that place and people, who is caring for the animals before experiments, and who is producing laboratory materials for study, among other questions about our relationality to all involved in making a particular experiment possible.

Some recent feminist science studies literature brings questions of invisible, racialized, and gendered forms of labor to the center of debates about the ethics of new technologies (Atanasoski and Vora 2015; Herzig and Subramaniam 2017). Earlier feminist science studies literature by explicitly Marxist and/or socialist feminists brought up labor questions. Hilary Rose (1992) suggests that modern sciences produced a split in what kind of labor counts as scientific production, relegating technicians (those who do the hands-on work of science) to non–knowledge producers. Their labor does not count as scientific work and is not counted in most publication practices nor included in professional associations.

These labor issues are important to reengage with as we challenge the definitions of the scientist. As feminist scientists think about movements to embrace hands-on work, we need to ask questions about labor throughout the process of scientific knowledge-making. When we say labs of our own, we cannot mean simply a laboratory run by an academic feminist who continues practices of ignoring who is doing the labor.

Additionally, the "our" of labs of our own must go beyond who operates and labors in the lab to include those whom the research impacts and all who impact the kind of research conducted. To this end, Barad (2001) argues for science literacy that acknowledges everyone's role in producing science. This proposition draws on the most basic science studies insight that all knowledge is culturally and historically situated, and therefore everyone/culture is part of science-making. Thus, to produce more ethical science, all of us must become more ethically involved in the process.

Who, then, is *our* in the framing of feminist labs of our own? Virginia Woolf's original claim for space of one's own and the economic means for women to be able to write has been critiqued not only for centering Western white women but for its reliance and complicity in colonization for these women to be successful (Zimmerman 2012). In addition, the framing of a room of one's own focuses on individual success. A focus on individual success misses the way that racial and gendered capitalism requires large groups of people to be disenfranchised, to work for others' profit, and to have differential life chances. That is, while individuals of an oppressed group may be able to move into the privileged group, the capitalist world system would not exist if everyone had the same access to resources. Collectivity has therefore been important in any movement that seeks to radically change systems of oppression and domination.

These politics come through even within feminist science studies scholar's interventions into mainstream science studies. For example, Rose (1992) finds the laboratory and academy to be sites of potential shifts in culture, but she critiques the claim by Bruno Latour (1983) regarding what an individual can do with a lab of their own in his oft-quoted defense of laboratory studies: "Give me a laboratory and I will raise the world." Instead, she suggests that feminists' collective action has produced spaces of their own in the academy, and these laboratories of our own could "change the world." Interestingly, Rose, like many others, read Woolf's original articulation as a collective call for space. However, the call for a universal womanhood has time and time again meant rights for a racially and class privileged group of women at the expense of other women, men, and queers.

Although the shift to collectivity over individualism is important, it is also not sufficient. We must be careful to not assume commonalities across difference in such a way that obscures material inequities. In the preface to Keller's tenth anniversary edition (1995) of *Reflections on Gender and Science*, she reflects on what "we" and "feminism" meant at the time the book was originally written compared with the time since. She suggests that the time of its first publication in the mid-1980s "appears to have been the political and emotional peak of our own particular revolution—at a time when the word feminism resounded with hope and promise, when its primary connotations were still those of emancipation, empowerment, and, perhaps especially, community" (xiii). She writes with what seems to be a sort of nostalgia for a time when "we" was used and was thought of as better than "I." She discusses important reasons for rejecting the individual for the collective, but there seems to be a subtle racism at play in her claim that that period of innocence is over. The innocence that has been lost has given way to a complication of gender as plural. She does not explicitly discuss the critiques and change as about race, but we can assume based on feminist history that this is part of what has happened. Although she affirms that these changes were necessary and are correct, making reference to simpler, more innocent times is often in U.S. popular culture a way to lament the end of slavery without explicit racism.[6]

Women of color feminist scholarship has worked to complicate identity, community, and solidarity. Throughout Grace Hong's book *Death beyond Disavowal* (2015a), she repeatedly goes back to a revised form of Audre Lorde's (1984) question, "In what way do I contribute to the subjugation of *any part of those who I call my people*?"[7] Lorde's partial answer to this question was, "Insight must illuminate the particulars of our lives: who labors to make the bread we waste, or the energy it takes to make nuclear poisons which will not biodegrade for one thousand years; or who goes blind assembling the microtransistors in our inexpensive calculators?" (1984, 139). Hong argues that Lorde points out differences within Black communities not to argue for more inclusion but rather to point out how those members of the community who have been offered (even precarious) inclusion are offered this at others' expense:

> In observing that inequities and forms of devaluation exist within and between racial groups, Lorde does not dismiss the brutal history of racial violence and devaluation that subtends racial categorization but rather calls attention to new configurations of power that, in contingently extending protection and value to formerly categorically marginalized identities, "are unevenly sutured to older categories of race, gender, and sexuality" [Melamed 2006]. Lorde provides a rigorous, deeply materialist analysis that extends beyond Black communities as well as the boundaries of the United States, providing a transnational analysis of the material costs of our own comfort. However, rather than allow this insight—that any politics of self-protection requires subjecting others to violence—to foreclose political possibility, Lorde instead crafts an alternative vision of a politics not based on self-interest. An alternative imagination of community that does not depend on identification or equivalence is neither easy nor utopian, for a truly relational vision of community must mean being willing to jeopardize one's own security for that of others. (2015a, 6)

Hong sets up a politics of self-critique against self-interest, importantly defining community ("my people") not through politics of universal, colorblind, genderblind politics but also not through a stagnant, essentializing definition of categories of difference either. What would it mean to consider our in laboratories of our own through this insight into the politics of difference in our neoliberal times?

To acknowledge the ways that borderlands are always spaces in transition, as Gloria Anzaldúa ([1987] 2012) defines them, we need to be as flexible as neoliberal gendered racial capitalism in understanding who are our people in need of laboratories of their own. To define *our*, we need to keep asking why we need these laboratories of our own. That is, why do we want to create our own knowledges? The *why* for feminist science studies scholars must be about changing the world, creating a more just world. This is not about individuals making it to become a CEO or president of a university or win a Nobel Prize or having more representation in biotech or science careers to make it more diverse and fairer; but instead it is challenging the ways that capitalism itself is racialized and gendered and how scientific knowledge helps to produce these truths.

One way we can produce new truths is by redefining and creating labs of our own. We must critique the continued patterns of gendered and racialized violence and discrimination that occur in the sciences at the same time as being aware of the ways that simply claiming inclusion or space for a minoritized group as an end goal will result in a redistribution of epistemic violence and discrimination without a dismantling of the systemic logic of inequality.

Conclusion

A queer science of failure would mean revaluing teaching and education and redefining science as not simply what is done in wet-labs. We will need to count all of us who "failed"—including those of us who dropped out of the pipeline in elementary school, those of us who never found our way to the pipeline to begin with, and those of us who had to hack our way out later downstream. That will mean working to make space in feminist spaces for those without PhDs in the sciences to be legitimate critical science literacy educators and producers of scientific knowledge. We must embrace an irreverent disdain of traditional sciences and practice feminist science by always keeping central the epistemic power that we are both challenging and risk reifying through our claims to do science.

We must keep analyses of power and accountability central to any engagements with Science, and exercise more care when we argue that we need laboratories of our own—asking *who* makes up *our* in this framing. One response to this question is to return to the old materialist call to center accountability to the most affected communities, to a material redistribution of wealth and resources as our aim of a redistribution of epistemic authority, in part through more democratic sciences. Drawing on the taxonomy of the Gesturing Towards Decolonial Futures collective (Stein et al. 2020), this work might move between "walking out" (defecting), "hacking" (redistributing university resources), and "hospicing" (actively participating in creating something transformative from the ruins of Western Science). Ultimately, combining feminist sciences in extra-academic spaces together with academic critical science literacy projects might yield a broader definition of feminist tinkering.

INTERLUDE 6

Queer Revolt

One week after I wrote the original outline for this book and one week before I entered Genspace's lab to start my ethnographic work, I attended a queer vegetarian Eid potluck. There among the group of anti-capitalist, anti-imperial queers, I met three science PhD students and recent PhDs who were fed up with Science. Does it matter that I met them at a queer vegetarian Eid potluck? It may, both because of the importance of identities and marginalization within the academy and because of the shared political orientation.

One of the students excitedly exclaimed at one point while we were chatting, "I just want them [scientists] to stop! I want to scream it!" and she was swinging her arms around and exasperatedly screaming it. A few moments later, with the same amount of passion and energy, she got out of her chair and bent down on the floor to use the wall as a prop as she explained the scientific experiments she was currently working on.

I started to tear up. I felt like I knew that complicated, conflicting emotional response she was having. Wait—it didn't feel complicated at the time. It was not so conflicting either. It was a surer feeling: I was more confident in the belief that we could make things better if we could educate scientists, if we could shut it down, if we could stop the bad things from happening. After all, we did not think scientists were bad—*we* were scientists—but we were angry that we had not been taught philosophy and history of science in our training. So we wanted to make it right.

She told me through our conversation that creating mandated education for scientists on scientific racism, women's studies, and other areas was her goal. This left me with lots of questions, which are now central to *Labs of Our Own*. I have questions about the role of passion for science for feminists and the role of feminists trained in the sciences. What is this passion for science that I saw in this student? Where does it come from? What does it mean? How do our feminist passions operate within the larger affective economies of mainstream science? These questions bring me back to asking how do we, could we, and/or should we as feminists

INTERLUDE 6   135

embody a passion for science? What kind of capacities to feel might we want to develop as feminist scientists and toward which objects and how?

The four of us decided that evening that we would start a radical queer feminist science group. The group decided that meeting in person and the relationships we built with each other were an important part of our project. Because I did not live in New York City, I was unable to continue—I only made it to one of the first meetings. This meeting with the four of us present convened in a conference room at the lab of one of our members at New York University. As we began talking, he told us that we were in the old Triangle Shirtwaist Factory building. As we looked out the new beautiful glass windows of the conference room we reflected on what it meant to be meeting in a space that was notorious as the location of an event that led to the strength of the U.S. labor movement. When a fire broke out in the building in 1911, the women and girls who worked at the Triangle Shirtwaist Factory could not escape because the owners had locked them in the workspace so that they could not take breaks or freely leave. Several of the workers jumped out the windows to their deaths in an attempt to escape the fire, and this horrific event was used to mobilize the labor movement to fight for better working conditions. And women as workers were centered in the fight.

We also learned that the building had another political history: its location in the Village made it a frequent place for gay men to meet up in its bathrooms. Glory holes, according to our host, could still be found in some of the bathrooms on the first floor. This led us to a discussion/debate over whether glory holes could be feminist, but we reached no conclusion.

We decided that the political significance and intersection of political movements made this the perfect place for the initiation of our group. For our group this represented our political orientation to the work. For us, the label that best explained that was queer. To us, queer signified politics that were interested in addressing root-cause, systemic oppressions as opposed to a representational politics looking to increase the number of gay people who are included in the upper ranks of an unjust capitalist system. This position has often been best explained in contrast to a gay mainstream rights agenda that "fights for an end to discrimination in housing, but not for the provision of housing; domestic partner health coverage but no universal health coverage . . . for tougher hate crimes laws [to address antiqueer violence], instead of fighting the racism, classism, transphobia (and homophobia) intrinsic to the criminal 'justice' system" (Bernstein Sycamore 2004, 2). We were committed to understanding and working to undo the power that Science had in defining our bodies, identities, and worlds. We were interested in making connections between labor politics, sexual politics, and antiracist politics.

In the next chapter I look at the intersections of various social justice movements and their place in the production of feminist science. I name this interlude "Queer Revolt" to mark an interest throughout this book in how feminists can challenge the status quo of Western Science. It marks for me a political orientation that is interested in disruption and in creating new possibilities, and is based in the idea that ethics and accountability are critical elements of a liberatory politics.

CHAPTER 6

Tinkering as a Feminist Praxis

My first introduction to democratic synthetic biology was through learning about the Pink Army Cooperative. Based on the name, I expected that this was a feminist collective, but I quickly found out that my assumption was flawed. The founder of this group was Andrew Hessel, and his board of directors was originally filled with male business and medical professionals. On his website, he used terms like "revolution" and "community-driven" to describe the project (see Figure 6.1 for a screenshot of its homepage). The actual cooperative structure remained unclear as he became a self-appointed spokesperson for the field of synthetic biology.

At that time, Deboleena Roy and I were working on an interdisciplinary ethics project on synthetic biology, and we invited him to speak to our class (Giordano 2016). Hessel's goal was to get people to pay $20 into his "cooperative," and his theory was that he could create a personalized therapy by producing genetically specific medicines to cure an individual who had cancer. The cooperative would support this experiment on one woman. In other words, he was crowdfunding before crowdfunding had really taken off. During our time together, he repeatedly asked me to give money to his project, and I repeatedly refused. He could not believe that I would not give up a mere $20, and I listed off at least ten other things I would rather give $20 to—letting him know that I would rather drop $20 in a hole to never be found again than give it to his project.

His shock at my refusal to support a "good cause" is a familiar problem for those of us who question the good causes for which our grocery stores collect our extra pennies or even our beloved queer communities who sometimes advertise events as benefiting "a good cause" without any specification of what the cause is. The assumption that charity and donations, particularly in science, are universally good makes one by some transitive property of social good a good person. Supporting and doing science is routinely cited as an unquestioned benefit to society despite decades of evidence from science studies scholars that show science to be a culturally situated practice.

Figure 6.1. Screenshot of Pink Army Cooperative homepage. The tagline for the browser reads, "Pink Army Cooperative: Join the Revolution!" The words "Pink Army" and symbols on the page are in two different shades of pink. Archive of http://pinkarmy.org captured by the Wayback Machine (https://web.archive.org/web/20111228043216/http://pinkarmy.org/) on December 28, 2011. Last accessed February 23, 2024.

During my conversation with Hessel I pressed him to explain why he chose breast cancer of all things to try out his idea, assuming there would be some personal connection. Instead, he replied that he chose breast cancer because there was already a movement that he thought he could tap into for funding. The unintentional co-optation that I trace in chapters 2 and 3 of this book was in this instance clearly articulated in the project plan. I was shocked. This project was my first clue that community biology projects may use democratic social justice rhetoric without really engaging social justice movements or deeper principles. It may also be the most direct co-optation of feminist and social justice language and movements and even an example of pinkwashing over other concerns with biotech.

But were there actual feminists engaged in synthetic biology and do-it-yourself (DIY) science? It seemed like a natural fit, but as Roy and I worked on our interdisciplinary project we could not find any self-identified feminist biohackers. Instead, we read other feminist activist and academic projects as potential projects in feminist tinkering, although the participants do not identify as tinkerers per se. We shared two of these activist projects with students. The first was sub-Rosa and the other a representative from the recently defunct Committee on Women, Population, and the Environment.

Our feminist science past and present are intimately entangled in mainstream sciences, democratic sciences, and broader political questions of democracy and capitalism. Feminist tinkering is part of a longer feminist engagement with science that begins before the new democratic sciences I describe in earlier chapters. *New democratic sciences* refers to the discourses formed around the DIY biology movement over the last decade and a half. The prefix "new" serves to contradict the idea that there have never been desires for and trials in democratic science before. Indeed, there have been many instantiations of democratic sciences—including feminist health and environmental justice grassroots science movements. These movements set the stage for newer democratic sciences as well as acknowledging that any of our current feminist embrace of tinkering is influenced and co-produced by this new moment of mainstream engagement with democratic science.

Over the last several years, some feminists have developed explicitly feminist biohacking groups. However, there has been scant attention to these groups, just as feminist hacking spaces (focused on technology instead of biology) have either been ignored or collapsed into a narrative of hacking as one monolithic community (Foster 2017; Maxigas 2012). The mainstream construction of the biohacker as male forecloses reading other potential female science projects as biohacking (Jen 2015). By documenting and analyzing hackers at the margins, scholars have aimed to show the multiplicity of hacker identities (Foster 2017) and increase the possibilities for a social justice–focused agenda to become more popular (Maxigas 2012). Although this work has focused mostly on computer and tech-related hacking, biohacking and activist engagement with biology can provide a basis for new genealogies of both feminist biohacking and mainstream biohacking.

From Armies of One-Breasted Women to Pink Armies or Genderf-ckers? Competing Genealogies for Feminist Science

Feminist and other activist health projects of the late twentieth century have set the stage for queer, anticapitalist, antiracist biohacking projects as well as neoliberal co-optations. Maxigas (2012) argues that the genealogies that individual computer hackerspaces most identify with influence the political orientation of the particular spaces. He traces two differing genealogies: one is an individualized, antiauthoritarian liberalism exemplified by the Steve Jobs garage story; the other is anticapitalist, autonomous (anarchist) movements circulating mostly between European and Latin American (and some North American) communities. Maxigas's work acknowledges the interplay between these genealogies at the same time as arguing that there is an attempt to erase the activist genealogy by collapsing the two kinds of spaces into one. Similarly, clearly delineating a queer, feminist, antiracist genealogy can help not only to illuminate the various biohacking communities that exist today but to highlight a way for feminists to build on these traditions as we engage with science and technology.

In the 1970s and 1980s feminists and other activists outside and then inside the academy pushed for control over science and health research, resources, and policy decisions. Groups across the spectrum of liberation movements were part of a broader radical health movement; for example, the Black Panther Party coordinated with other groups such as the Young Lords and feminist groups such as National Organization for Women as part of their health activism (Nelson 2011). Movements that worked together to point to racism in scientific theories and the neglect of the health of Black and Brown communities can be seen as part of what Alondra Nelson (2011) calls the "long medical civil rights movement" and as precursors to current day environmental justice movements that coined and made visible environmental racism (Bullard 1993). These earlier health movements also provided lessons for 1980s and 90s HIV/AIDS activism (Epstein 1996).

These health movements came with a certain in-your-face affect and politics against silence. Take, for example, Audre Lorde's rhetorical question, "What would happen if an army of one-breasted women descended on Congress and demanded that the use of carcinogenic fat-stored hormones in beef-feed be outlawed?" (Lorde 1997, 16). Originally published in 1980, Audre Lorde's *Cancer Journals* represents a certain kind of health activism that is based in personal testimonies, materiality, and collective action. Modern day pink ribbon campaigns capitalized on these earlier activist movements, dominating discourses on breast cancer (King 2006). However, the story is more than a simple progressive history from an angry feminist movement to a positive smiling pink ribbon awareness campaign. There has also been ongoing activism that refuses cooperation with multinational corporate interests.

The move toward more democratic sciences that has been realized through DIY biology community laboratories and tinkerers has also drawn on histories of activist health movements. Michelle Murphy asserts that, "Feminist self help, with vaginal self-exam as its iconic protocol, was one of the most sustained efforts to practice science as feminism. It was also a reassembly of objectivity that reverberates in participatory methodologies of many political stripes today" (2012, 100). Mainstream DIY biology, as I discuss in earlier chapters, co-opted such social justice–based participatory methods through a "political stripe" more interested in entrepreneurialism than social justice. However, the story is not simply about co-optation. Murphy points out in her analysis of the radical feminist self-help movement in California that this movement was entangled itself with an emerging neoliberalism, where activists found themselves both resisting and co-producing a new racialized, economic order. That is, the idea of a pure feminist, antiracist, social justice–oriented origin to these movements is not accurate either. There were and are competing political ideologies within self-described feminist and social justice movements.

The genealogies in which a group situates itself can matter for how it operates and what it produces (Foster 2017; Maxigas 2012). There are multiple definitions and genealogies of feminism as a political standpoint itself. A mainstream definition of feminism relies on progressive narratives of inclusion of women in

Western society. In this account, gender as we know it is assumed as always existing; both history and cultural existence before the sixteenth century are flattened or nonexistent. Assumptions of capitalism as part of the progressive history of mankind are embedded in this view in such a way that being included in the current dominant global system of capitalism is seen as the only way to create a more just world. The pink ribbon machine is an example of this kind of feminist maneuvering; it naturalizes and protects capitalism by reinforcing gender and sexual norms even while including certain disabled bodies (women with breast cancer who "survive"). Pointing to the systemic exclusion of large groups of society is critical for challenging gendered racial capitalism because it operates by forming these very groups of difference to justify a highly stratified society.

Challenging this system of exclusions cannot be simply accomplished by greater inclusion of the excluded groups into the same global capitalist system. The mainstream DIY biology labs that I followed relied on romantic genealogies of computer mavericks such as Steve Jobs and ahistorical, untainted science exploration before Big Bio and bureaucratic institutions became involved. A feminist practice that simply seeks to be included in a mainstream DIY biology movement by increasing women or minorities in those spaces (or even creating exclusive women-only spaces) may overidentify with this neoliberal genealogy that co-opts the language of revolution, justice, and collectivity from movements of the 1960s and 1970s, leaving the deeper politics aside and instead focusing on an individual economic subject above all else. This is not to say that intentionality around *who* is invited into these spaces is not important. But inclusion around one axis of difference, such as gender, as if it is not related to the larger social-economic system of gendered racial capitalism is an insufficient condition for building radically new worlds based in social justice. Feminist biohacker groups who identify with anti-capitalist genealogies provide a point of divergence from the current mainstream biohacker genealogies.

The group Quimera Rosa, originally founded in Barcelona, Spain, in 2008, identifies as transfeminist artists and biohackers. When I asked the group members to explain the movements that they see themselves as a part of, they listed post-porn and transfeminism movements. They described the two as interrelated through the centering of encounters with gender, identity, sexuality, and night politics. They used the term *encounters* multiple times as a way to think about creating new things and new possibilities in different moments and in contrast to a liberal model of inclusion. That is, they did not see their activism as simply making visible more identities so that everyone is eventually included and protected through legal frameworks. Instead, they see their activism and work as about opening new ways of thinking and being. They highlighted the idea of these movements being formed by networks instead of assemblies so that it was about interconnectedness and not trying to come to a universal consensus for the movement.

Importantly, they pointed out the way they were using "transfeminism" was different than some of the articulations of transfeminism in U.S. contexts. Transfeminism to them is not about including trans* people into women's movements. Instead, they see it as a way to mark a feminism that challenges such categories as gender and sex altogether while they tinker with those identities to produce a multiplicity of genders and sexualities. Although they draw on and integrate some U.S. and British queer theories and theorists, the Spanish and French transfeminist movements have been characterized additionally as a challenge to an English and white centric queer theory (Espineira and Bourcier 2016). Karine Espineira and Sam Bourcier (2016) argue that transfeminism, with its incorporation of post-porn politics, works not only with "resignification but also rematerialization" (84). That is, transfeminism works with bodies and spaces as it challenges categories of identification.

As I was frequently reminded by the transfeminist activists I interviewed, U.S. gay and trans-identity-based movements are not good comparisons for the political positioning of their work. Reproductive justice and environmental justice movements and particularly the intersections of these movements might be a better analogy to the Barcelona-based transfeminist, post-porn movements that they discussed. Those at the intersections of those movements bring to the forefront feminism and queer politics. Similar to the movements that Quimera Rosa described in Barcelona, a reproductive justice movement that has formed through networks across the United States and Canada incorporates a pro-sex politics and focuses on an intersection of issues instead of increasing rights for one group of people.[1] Disability and health justice have also become increasingly central parts of this movement, and the intersection with environmental justice has been instrumental in identifying health for communities at all stages of development as important, moving beyond a narrow focus on questions of abortion and pregnancy.

This overlap between reproductive justice in the United States and post-porn/ transfeminism in Spain is evident in the way Gynepunk, the first self-identified feminist DIY biology group, explain their project and political orientation by centering a critique of J. Marion Sims. Sims, who is often referred to as the "father of gynecology," has been critiqued by academics and activists for his use of enslaved women in experiments, sometimes without anesthesia (Washington 2006); his work exemplifies the unequal benefits and harms of research based on racism. The U.S.-based reproductive justice movement centers the lives and needs of women (and queers in newer iterations of reproductive justice definitions) of color as a way to get to the root of capitalist and colonial control of women's bodies through systems of race and health. Therefore, Gynepunk's work can be seen as an intentional desire to link to this kind of politics.

As a note of caution against oversimplifying these movements, the language of *reproductive justice* has been taken up by a broader segment of the reproductive rights movement over the last several years. In some ways this has brought a deeper analysis to some mainstream organizations, but it has also meant that the battle over defining reproductive justice and its principles has become more contested, and in some cases the idea has been used in name alone without challenging middle-class, white-based reproductive rights agendas.

Many feminist scholars trained as scientists became critical of dominant sciences through activism outside of the academy. For example, Jan Clarke describes how she came to a feminist approach to science through activism in the labor and women's movements: "As I became more politically active—involved in union work and local women's movement activities—my paid work in biology research seemed isolated from people and issues around me. To resolve the contradiction, I found a way to shift from lab research work to government funded research projects that required my science education to interview scientists, engineers, and tradespeople in order to develop questionnaires and write occupational research reports" (2001, 36). My own transition to feminist science had to do with my activism outside of the academy as well (Giordano 2014), particularly with reproductive justice activists. Deboleena Roy's (2004) research was also shaped by her involvement in reproductive justice movement work.

In the introduction to the *Feminist Science Studies: A New Generation* (2001) reader, Maralee Mayberry, Banu Subramaniam, and Lisa Weasel explain that the chapters in the final section "illustrate the rich potential that feminist science studies brings to questions relating to community and activism" (8). In one of these chapters, Lisa Weasel (2001) puts forth a proposal for the development of feminist science shops taking the Dutch model of science shops in which extra-academic communities set research agendas for graduate students. An additional site for early feminist science is Jean Barr and Lynda Birke's *Common Science?* (1998). Throughout the text they argue for centering activist communities (e.g., pp. 6, 126). They also refer to the science shop as a potential model for a more democratic feminist science (138).

In both texts on doing feminist science they argue that feminist science must follow suit with other kinds of feminist projects that work across academic boundaries and the theory–practice divide (Barr and Birke 1998; Weasel 2001). In my own work I have taken up Weasel's suggestion to incorporate this into our classrooms (Weasel, 2001, 317). Activists have made new sciences and been made new themselves through their engagements with health and science.

Activists Producing New Scientific Knowledges, Collective Identities, and Worlds

Many kinds of activist interventions fall under the umbrella of a *feminist tinkering methodology*. Challenges to the myth of a pure scientific objectivity have taken place both in the streets and in academia, led by professionally trained scientists, nurses, doctors, and philosophers along with laypersons who became self-taught or community trained. The challenge was twofold: dismantling specific racist, sexist, and homophobic scientific assumptions and calling for new scientific research based in community need and knowledge. These challenges called attention to the impossibility of absolute neutrality (Harding 2004) while putting forth better models for producing knowledge—sometimes called feminist objectivity (Haraway 1988).

Taken together, these activists' work shows that critiques of science and *doing* science need not be held apart.

Health activists in the last half of the twentieth century were not simply interested in fighting for more access to health resources, nor were they simply critiquing technosciences. Rather, they were producing new scientific knowledges. For example, Murphy (2012) claims that the self-help movement she traces represents "attempts *to do* feminist technoscience" (21, emphasis in original). The vaginal self-examination can be seen as a challenge to scientific objectivity itself, thereby valuing a different way of producing knowledge (Murphy 2012). Another example is U.S. disability activist tinkering, which Aimi Hamraie (2017) articulates as "knowing-making," highlighting the inseparability of the two.

Likewise, present-day activist interventions also represent *doing* science itself. This doing is politically situated, always acting from an understanding of culture and science as inextricably linked. Scholars across critical studies fields such as Black studies, feminist science studies, Indigenous studies, disability studies, decolonial studies, and Chicana/o/x studies have argued for the need to produce new sciences. Sylvia Wynter's (1994b) call for "rewriting" knowledge recognizes the responsibility the humanities and sciences bear for a co-produced ("written") difference. The goal of rewriting knowledge (producing new sciences) is explicitly linked to a political objective, that of undoing the disciplinary boundaries and truth that produce "narratively condemned status[es]" (70) of Others/nonhuman humans. Matthew Weinstein (2010) offers the language of critical science literacy to read activist engagements with science and particularly health in which activists use already existing scientific knowledges and produce new scientific knowledges "to help achieve radical visions of social justice" (267). Activists rewrite knowledge through various forms of critical science literacy.

Methodologies, Processes, and Protocols

Feminist philosophers and science studies scholars have argued that epistemologies, methodologies, and ontologies are inseparable from each other and politics. Tinkering as a feminist science methodology takes as a starting assumption that how we do research is connected to how we understand what we know, and what is possible to be known.

Faith Wilding and Hyla Willis lead the cyberfeminist art collective subRosa. In 2011, as Wilding and Willis prepared to give a lecture at Emory University, Roy and I as their hosts asked if they needed anything for setup. "Can you find us lab coats and Q-tips?" they asked. Also, they asked us to limit access to the lecture hall to one door only. As we guarded the second door, students and faculty lined up outside the other door to gain entry. Inside the door, our guests (subRosa) in their lab coats (which we borrowed from friends on campus) handed each participant a Q-tip and a sheet of paper with an empty box and several questions. subRosa instructed the participants to swab their cheek and wipe it on the box on the paper and then fill out the questionnaire. This was a rather strange and unexpected welcome to what was billed as a colloquium event.

Nonetheless, most of the line complied and sat down with their questionnaire to answer questions about themselves such as their eating habits and religious preferences so that the "researchers" could see how these characteristics might be linked to their genetics. A few protested, "I am not giving you my DNA." subRosa calmly replied, as they made notes on their own paper, "So you are refusing to participate in this data collection?" They explained it was perfectly acceptable to opt out, but they just needed to make a note of it.

I watched, amazed at the performance that was taking place. What was it that made everyone participate? Even those who opted out participated in this performance, not questioning whether subRosa could actually sequence DNA from a cheek swab left on a piece of paper or whether that would tell them anything about the answers that they asked in the questionnaire. Was it the authority of the lab coats? The trust in us as the organizers? The lack of knowledge about DNA collection?

The lack of options to challenge genetic research became clear through this exercise. The pressure to participate and the option to refuse through opting out demonstrated a limitation to the idea of consent in genetic research. That is, opting out is not always an easy option, and it does not necessarily stop the harm to those opting out because of the continued power of genomics on defining who we are. Through a playful, creative, and artistic approach, this vignette demonstrated that we need to rethink what kind of ethics are needed beyond individual consent (Benjamin 2016).

A few decades earlier in the 1970s, U.S.-based feminists would wear white laboratory coats to purchase pregnancy tests that at that time were only available to doctors. This role-play was welcomed by the businesses, who did not care who they sold kits to (Murphy 2012, 27). The "labs" of their own (the feminists of the women's health movement), however, did not require lab coats, nor did these "clinics" take place in sterile laboratories at all. In fact, the feminist self-help clinic represented the name of events and not places (Murphy 2012, 26). Murphy argues that the Self-Help Clinic "was a mobile set of practices, a mode for arranging knowledge production and health care . . . a protocol," which she explains "establishes 'how to' do something, how to compose the technologies, subjects, exchanges, affects, processes, and so on that make up a moment of health care practice" (2012, 25–26). Drawing on Gilles Deleuze and Félix Guattari's concept of assembly (1987), Murphy highlights the importance of the protocol as defined by the specific way that materials and activists are arranged in a particular moment, identifying it as a dynamic process rather than something that can be explained by a simple summation of its parts. She further argues that these protocols were formed through an "internal ethic of *flexible and experimental* reassembly" in which activists aimed to share knowledge, demystify biomedical knowledge and *tinker* with it to repurpose it for their own political goals. The tinkering was constant in that it could be repurposed by different communities, in different moments. The process of sharing and making self-help clinics was about constant iteration. Part of the methodology was also about *hands-on* knowledge sharing. The "self" in self-help was about experimenting and learning on real bodies in real situations instead of in sterile doctor's offices. However, importantly for this early instantiation of feminist tinkering, "self" did not mean an individualistic activity but was rather always about collectivity—in experimenting and knowledge sharing (Murphy 2012, 26–27).

The kind of consciousness-raising from 1970s feminist activism can be seen in the popular methods that continue to be incorporated into activist and community-based knowledge production such as personal narratives and storytelling. Lorde's *Cancer Journals* represents this popular strain of recording, sharing, and politicizing personal narratives, which operates at the intersections of feminist and disability studies (Garland-Thomson 2005). Narratives focus on knowledge from specific individuals, but the goal is to speak out to connect with others with similar experiences to highlight the systemic nature of disabling narratives that propagate through the medical model.

In a storytelling project called *Depo Diaries*, the now defunct Committee on Women, Population, and the Environment (CWPE) collected firsthand narratives of people who had taken Depo-Provera birth control. The project included a way for people to send in their stories long distance, thereby connecting these stories not only across the United States but also from other locations around the world. Their method of collecting data was not simply part of the effort to find out the truth about side effects. Through the work of listening to stories or reading them and then sharing stories with others, CWPE was organizing. Therefore, the practice of knowledge production and sharing was also about producing a shared political agenda and was foregrounded by their reproductive justice and environmental justice frameworks. This introduces a different kind of ethics that is dissimilar from the focus on clinical trials. This ethics goes beyond the idea of informed consent and the autonomous subject while focusing on individual stories. It has to do with political knowledge building with a focus on community health and systemic issues.

This form of knowledge production can be understood through feminist standpoint theories. In fact, the women's health movement could be seen as developing a case for the formal articulations of feminist standpoint theories that were not made until the 1980s and 1990s in academic literature (Murphy 2012). Both *Our Bodies, Ourselves* (Boston Women's Health Book Collective, 2011) and the lesser known but similar text *A New View of a Woman's Body* (Federation of Feminist Women's Health Centers, 1991) bring attention to the connections between *who* is producing knowledge and what kind of knowledge is produced (Davis 2007). The title *A New View of a Woman's Body* invokes the idea that there are different views from different positions. This framing fits with Donna Haraway's argument for situated knowledges as part of a feminist objectivity: "I am arguing for politics and epistemologies of location, positioning, and situating, where partiality and not universality is the condition of being heard to make rational knowledge claims. These are claims on people's lives; the view from a body, always a complex, contradictory, structuring and structured body, versus the view from above, from nowhere, from simplicity" (1988, 589). The illustrations in *A New View of a Woman's Body* intentionally locate the observers and their apparatuses within the image. There are illustrations with women's hands and speculums included, and the viewpoint is most often looking at the body from the woman's own eyes. This contrasts with many medical illustrations that are from the doctor or researcher's viewpoint, looking at the woman and giving a sense of a neutral observer. The authors of *A New View of a Woman's Body* saw their text and illustrations as demonstrating a diversity of women's bodies—considering race in particular, as well as a wider range of body shapes and sizes than standardized medical texts.

Standpoint theory was also important for Black liberation movements. The Black Panther Party's challenges to scientific objectivity were likely influenced by their commitment to communist ideologies. Standpoint epistemologies can be traced to Marxist theories that argue that the proletariat has an opportunity to develop a more accurate viewpoint of the world due to their hands-on labor and the multiple viewpoints they can see from due to their oppression (Hartsock 2004). The Black Panther Party held central an idea that their communities were better positioned to understand and treat the problems they faced than outsiders.

Disability activists and scholars also make claim to be better positioned to produce knowledge about disability. Hamraie (2017) argues that the key intervention of the social model of disability is often misidentified as the argument that disability is environmentally produced. Hamraie shows that the emergence of the social model of disability by activists was in opposition to not only a strict medical model but also the models put forth by rehabilitation sciences. At that time, rehabilitation science was also focused on environmental constraints and removing them to make sure that disabled people could be productive citizens. Hamraie argues instead that while disability activists in the 1970s "agreed that disability is a socially and architecturally produced disadvantage, [they] asserted that their lived experiences made them better experts on the subjects of disability and challenged the norm of compulsory productive citizenship" (12). Hamraie therefore argues that this challenge was about making claims for a "new disability epistemology." This way of knowing from the lived experiences of disabled people represents another example of the way standpoint (or sitpoint) theories were critical to the mid-twentieth century activism in the United States.[2]

Another key method for feminist activists and academics has been engaging directly with scientific literatures through critique. Teaching feminists to read and engage with scientific literature can itself produce new scientific knowledge (Giordano 2017a). This is based in part on the understanding of our collective roles through disbursed modes of power/knowledge in producing and maintaining scientific truths. I call this practice in my classroom a *feminist critical science literacy*. Arguing against the racist and sexist underpinnings of science was also a key part of 1970s and 1980s social movements. Scientific literacy allowed the Black Panther Party and its communities to use scientific theories about human difference for politically strategic goals. "For example, with its sickle cell anemia campaign, the Party repurposed evolutionary theory to argue that this genetic disease was an embodied vestige of slavery and colonialism. At other times, the activists rejected biological theories about race and the 'nature' of communities of color as they did in their campaign to put down UCLA's planned 'violence center'" (Nelson, 2011, 20–21). These programs constituted a kind of tinkering with science through the Black Panther Party's reassembly of biological theories.

Reassembly, of course, can continue to take place and be initiated by other actors. In the case of sickle cell anemia, the reassembly by the medical establishment and government resulted in mandatory screening laws being passed that then resulted in racial discrimination in hiring practices (Hubbard and Wald 1999). Discrimination took place specifically in terms of hiring pilots and flight attendants. Although having sickle cell anemia can cause a danger in flying at high altitudes, merely testing positive for the trait does not. Hiring practices based on the presence of the trait thereby disproportionately impacted Black people. In addition, these mandatory screening programs were accused of promoting eugenic policies aimed at minimizing the number of Black people giving birth for fear of passing on sickle cell anemia. Therefore, activists had to always be politically conscious about their engagements with science and health. Presenting science as neutral or objective was not their orientation to this work. Their goals were about building political power and community control through contextualizing these scientific engagements.

Community Control of Knowledge Production

In mainstream science literacy campaigns the goal of demystification has to do with increasing buy-in to science with an aim for a specific affective relationship to science—one of science appreciation (Bauer 2009; Lewenstein 1992). Although demystification of science comes up in the case of the feminist health movement and Black Panther Party's work, the sharing of knowledge and production of knowledge is instead centered on community control and a related concept of self-determination. These activist traditions represent a cautious and ambivalent affective relationship to science where it is used more strategically for differing political purposes.

The Black Panther Party drew on Marxist-Leninist-Maoist theories that had an impact on how they thought about health. They saw this as a distinct trajectory from that of the U.S. federal government's health programs. Alondra Nelson (2011) uses the term "social health" to describe the Black Panther Party's approach to "health and biomedicine as being oriented by an outlook on well-being that scaled from the individual, corporeal body to the body politic in such a way that therapeutic matters were inextricably articulated to social justice ones" (11). Bobby Seale and Huey P. Newton met in a community action program run by the federal government as part of the "war on poverty," but they reappropriated (or reassembled) it into "serve the people" programs. These programs were "instituted as a more democratic and participatory alternative to federal ones" (Nelson 2011, 72).

Similarly, the feminist women's health movements often used pre-existing medical knowledge at the same time as reconfiguring it with their own experiences. As exemplified by the title and slogan, *Our Bodies, Ourselves*, ownership and control over the knowledge was central to the movement's goals (Davis 2007). Importantly, in each case, these were dynamic systems of knowledge that were iteratively produced and edited over time. *Our Bodies, Ourselves* was not simply a book volume but a living document. An important part of the project was receiving letters from readers and responding publicly to those letters, integrating more voices into future editions of the text.

Nelson notes, "In keeping with the era's DIY spirit, the [Black Panther Party] activists enacted the better world they imagined by establishing their own independent healthcare initiatives and institutions" (81). What is particularly important in the DIY spirit and movement for democratic science that she situates the Black Panther Party's activism within was the focus on community control. DIY did not mean doing it alone in one's home, similar to how "self" help did not mean independent, isolated medical care. Instead, DIY here was about creating health institutions that were controlled by Black communities. This was articulated as part of an agenda of self-determination. This self-determination at a community level sought to challenge the very way people lived, how gender and sexuality were thought of, which therefore meant experiments in new identity formations at the individual and community levels (Spencer 2016).

Reconfiguring Identities and Coalition Building

Feminists collectively tinkering have the ability to produce new kinds of collective identities to challenge biological truths and thereby political truths. The tinkerer emerged as a particular kind of subject as community biology labs were formed. This identity formation of "the tinkerer" happens through association with democratic science, which merges ideas of a colorblind and genderblind total inclusion and the idea of science as pure and innocent. The idea of the tinkerer as a type of person who is perfectly positioned (born?) for this kind of scientific exploration is proven through a passion for science. Integral to this formation was that the tinkerer was systematically hampered from doing the kind of science he wants to because of traditional sciences. This positions the tinkerer then as a scientific explorer who is disenfranchised and in need of space of one's own. When tinkerers form community, this identity forms into a particular political and ethical community: the "proper informed public." This particular public holds science and democracy as inevitably good, and by adhering to the supposed universal subject of the tinkerer, the racialized and gendered boundaries of this community were obscured. This public holds the claim of being the most legitimate stakeholder in debates. What then does a collective feminist tinkerer identity produce?

Understanding feminist tinkerers' political identities and their relationship with more mainstream tinkerer identities as well as with other feminist, antiracist, anticapitalist, and decolonial social movements is important for developing a feminist engagement with science that challenges rather than supports racial capitalism. These movements have been part of producing individual and collective identities. For example, Murphy (2012) explains how the feminist women's health movement shifted slightly who was included in those with the potential to be liberal, self-realized human subjects, so their work also strengthened overall the idea of liberal individuated persons.

Importantly, as Murphy notes, this collective formation of "sisterhood" imagined itself as a universal collective of women that was unraced and unclassed. This idea of the universalized women subject by a group that was primarily made up of middle-class white women was challenged by feminists of color. At the same time as the universal idea of womanhood was propagated throughout some parts of the feminist health movements of the 1970s and 1980s, other groups of feminists led by feminists of color were articulating the idea of what we now generally call "intersectionality." For example, the Women's Community Health Center of Cambridge argued that oppressions based on gender, sex, class, and race were "intertwingled" while the Combahee River Collective's Statement, which highlighted the intersections of capitalism, sexism, racism, and homophobia, also was dealing centrally with health projects (Murphy 2012). This continuing challenge to the unraced and unclassed category of "women" can be seen in the articulation of reproductive justice in opposition to the idea of individual choice as the foundation of a feminist politics.

In disability activist movements of the 1970s and 1980s, collective identities were also formed together through new knowledge-making politics. "Similar to the feminist women's health movement driven by texts such as *Our Bodies, Ourselves*, disabled people organized knowledge and expertise around their independence from medical authority and interdependence with one another" (Hamraie 2017, 112). Creating a sense of community requires not only a positive affirmation of who belongs but an oppositional statement of who the community is not. The formation of a collective identity had to do with a distancing from medical authority (independence) and a collective practice of knowledge-making (interdependence).

The formation of identity politics-based organizing in the 1960s and 1970s can also be seen in the example of the Black Panther Party. In the case of the Black Panther Party and their involvement in medicine and science, their health politics did not seem to reinforce the individual body in the same ways as the idea of self-help groups; however, they did argue for a community-based self-determination. Ideas of well-being were always linked to access to material resources such as food and shelter, which were systematically denied to Black communities in the United States. Many founding members had experience organizing against poverty from within government programs as well as being a part of student organizing that focused on not only access to education but access to critical education as opposed to education based in coloniality (Murch 2010). The Black Panther Party was able to articulate the stakes of controlling knowledge and its link to power and material conditions.

Although these various movements used identity politics to build community power and challenge the privileged position of white, property-owning males, they were also open to co-optation that reinforced racial capitalism by increasing diversity in government and corporate positions (Wynter 2003). This was despite some advanced understandings of identity from within the movements themselves. For example, the Black Panther Party carefully negotiated medicine and science both by challenging the abuses of medicine and science and by demanding access to health care. The incorporation of their demands for access was in part used as the rationale for a new wave of inclusion and difference studies in medicine beginning in the 1990s (Nelson 2011). Other largely identity-based movements such as the women's health movement, which had argued for more attention to specific health needs of women, and the HIV/AIDS movements of the 1990s, which had argued for gaining access to clinical trials, were also precursors to a renewal of research focused on measuring differences between groups of humans. The intersections of these movements' demands for inclusion and control were diluted and co-opted into the congressional response of mandating the inclusion of women and racial minorities in health studies along with a comparison of data across these groups. This inclusion–difference paradigm ironically set the stage for a renaturalization of difference (Epstein 2008).

In line with Wynter's (2003) call for a new science, feminist tinkering with science might provide one attempt to rupture the idea of the human as *Homo economicus*. Transfeminist, post-porn biohackers work to proliferate identities, not to try to capture all identities for inclusion but as a way to think about sex and gender as always fluid and changing possibilities for interrelationality. They are responding to the co-optation of previous movements and iteratively trying a new technique. In doing so, they are opening up new possibilities a feminist tinkering.

Toward a Feminist Tinkering: Decolonial Love, Affective Solidarity, and Ethical Practices of Knowledge Production

Despite its masculinist overtones, in many ways the idea of tinkering seems like a good fit for feminists interested in engaging more directly with science. First, there is a current association of "tinkering" with more democratic forms of scientific knowledge production, which feminists have long been interested in. Second, feminists have already theorized with many of the traditional and colloquial definitions of "tinkering" as a verb (e.g., playfulness). What, then, makes feminist tinkering with science different than the kind of tinkering being mobilized by biohackers and community bio labs? And why might it be important to distinguish? The short answer is an explicit political orientation to tinkering that centers an analysis of power and a goal of greater social justice—a kind of ethics to tinkering. The popular use of biological tinkering today highlights playfulness, hands-on work, and iterative experimentation. How have feminists interpreted each of these concepts in their teaching and their research methods?

Accountability is central to a feminist ethics. To be accountable to those most harmed by science's power, we must look beyond science studies, which has remained overrepresented by whiteness and heteronormativity in terms of scholars' demographics and its scholarship. One way these ways of thinking are reproduced is through citational practices: that is, who is cited and credited for key ideas and which ideas are thought of as theoretical foundations to build on for the field. Therefore, I engage here with an experiment in pulling together of theorists from feminist science studies with feminist, decolonial, and/or queer theorists who deal with ideas of knowing and being but not explicitly in conversation with science. If knowing and being can be thought of as the basis of scientific inquiry, then looking for resources for understanding ourselves and our worlds outside the proper science-related fields may help us to redistribute consolidated epistemic authority.

Several feminist science studies scholars have theorized the idea of becoming "response-able" knowledge producers (e.g., Barad 2001; Haraway 2008).[3] This highlights the centrality of ethical engagement in feminist science as a kind of practice, a way of doing science. In a 2004 essay, feminist scientist Deboleena Roy ends by answering a question she had posed, "Are we ready for feminist theory in science to influence the production of scientific knowledge?" (275). "The answer is yes, if we are willing. If we are willing to accept and recognize the limitations of our own work without letting these limitations stop us from doing science, or for that matter from creating feminist science. . . . And so to the feminist scientist working at the lab bench I say, READY. . . . SET. . . . PROCEED WITH CAUTION!" (277). The caution signals a kind of humility for the individual practitioner as well as an acknowledgment of the need to listen—to be part of collective action and collective reflection. Humility can also be a kind of modesty. The difference in this form of modest feminist inquiry compared to the archetype of the scientist found in the modest witness of traditional Western sciences is that it is not used to obscure accountability and power through a mythological objectivity. A modest feminist inquiry takes responsibility for one's role in the experiment and the larger world. The modesty is about understanding the important role of specificity and situated knowledges such that all vision is necessarily partial. This always places the observer in the experiment as opposed to the modest witness who is positioned to merely observe the truth of nature from their supposedly detached position. At the same time as there is a humility and modesty in what one understands that one could ever know, there is also a clear call for action. The need for taking feminist action is central to even a cautious, modest research.

In the current mainstream iteration of tinkering, playfulness and fun are deeply embedded in the practice. This fits into the entrepreneurial model of innovation. Acknowledging the current use of tinkering and democratic science for neoliberal entrepreneurial goals as a part of the context is a way of proceeding in this activity with caution! There is a danger in assuming that tinkering or any aspect of it (e.g., hands-on practice) in and of itself creates a better, more just science practice. As feminist methodologists have argued, a method is not feminist—or not in and of itself; it depends on how and why it is being used (Roy 2004). This assertion is based on the observation that the scientific method conflates methods, methodologies, and epistemologies all into one, where methods are "techniques for gathering evidence," methodologies are approaches to using various methods together in the interest of knowledge production, and epistemologies are theories of what knowledge and truth is itself (Harding 1987). Therefore, we can (and should? must?) imagine feminist sciences through feminist methodologies and feminist epistemologies without conflating these with methods, which are techniques that can often be used in both feminist and nonfeminist ways.

Sandra Harding suggests that behind the search for a "distinctive feminist method" is really the question "What is it that makes some of the most influential feminist-inspired biological and social science research of recent years so powerful?" (1987, 19). Forty years later, we are still trying to answer this question with a lot more research being claimed as feminist. We are playing with Harding's question: *What is it that makes feminist biohacking powerful?*

FEMINISTS' PLAYFULNESS

Play has become a central facet of the DIY biology tinkerer movement, and it has also figured prominently in feminist theorizing about methodologies and epistemologies. The ideas of play, living, and loving have been central to many feminist and queer activist movements. This has been posed often as a challenge to masculinist movements that continue to pretend that the public and private are separable, arguing against including cultural politics or personal politics in "serious" struggles. This idea of needing to live fully, have fun, and socialize with those you are organizing with appears in the many versions of the quote attributed to Emma Goldman, that if she can't dance, it's not her revolution. Although this phrase doesn't appear in Emma Goldman's writings, it likely was derived from her retelling of how a male comrade criticized her dancing during a party because her "frivolity would only hurt the Cause" (Goldman 2011, 56). This story is embedded in a book that documents several feminist frustrations with the masculine-oriented anarchist movement that she helped to lead.

One transfeminist activist of oncogrrrls when asked what brought them together with others, responded, "We are a bunch of people who like to play a lot." The play for these transfeminist, post-porn activists (including Quimera Rosa) in part takes place through what they called "night politics." Night politics includes drag performances that can be seen as a playing with gender, and BDSM, which has been described as playing with power. This kind of playing with bodies, pleasure and pain, and sex and gender through night politics was explained to be about building relationships and networks. These engagements with bodies and politics have opened up space for questions of health, able-bodiedness, disability, and disease.

By comparison, in the United States, reproductive justice fundraisers have included body painting, drag and burlesque performances, or even someone naked adorning the dessert table. I do not mean to suggest that *every* reproductive justice group holds such events. The decentralized nature of reproductive justice networks in the United States mean that there are local differences in how each organization takes up these politics. However, the centering of queer, trans, and disability politics seems to be consistent even if not expressed through sexual themed parties.

In each of these national contexts the introduction of art seems to open up more possibilities for playing with meanings, especially categories of difference, and for bringing up ethical questions. For example, the subRosa collective's demonstration that I described earlier playfully acted out DNA collection to raise ethical questions. During their workshop, Quimera Rosa talked us through the steps of a scientific protocol for testing a chlorophyll precursor, aminolevulinic acid, for its potential benefits in photodynamic therapies aimed at treating tumors. Their goal in moving through the steps of the protocol was to open up questions about the ethics of animal research, the relationship between humans and animals, transspecies entanglements, the (im)possibilities for ethical research within capitalism, and the (im)possibilities of working outside of capitalism, among other questions. This can be seen as the kind of "subversive repetition" that Roy (2012) suggests based on Patti Lather's (2007) articulation of this as a method of "getting lost," of "working the ruins."

During the workshop, when concerns came up and tensions rose about animal research, self-experimentation, and the visual imagery of needles and blood, Quimera Rosa suggested creating a Jar of Concerns. This was a physical glass jar to add notes to so that there would be a record of what had come up. To them this was a way to continue moving forward with their experimental workshop and to acknowledge any discomforts that arose. The jar did not mean that the topics were off limits for discussion. However, it was a physical reminder in the middle of the table that there were unresolved issues. Once the jar was placed on the table, the participants did not quickly move to fill it. It was unclear to me whether this was because it was an unsatisfactory response or because the presence of the jar itself acknowledged the continued concerns. Over the next few days, we did add some of the concerns into the jar. However, there was no time or process made to go over the concerns in the jar again.

Therefore, playing for the transfeminist activists who I interviewed was not only about having fun or having sex, although those are important parts of movement building and the political theorization process as well for them. Play in the sense of trying new things without predetermined end goals, a willingness to "get lost" in research and art and embrace the uncertainty of their experiments, was also central to their approach to DIY art/science.

In the classroom, it is a bit more difficult to be playful, perhaps. There are time constraints, space constraints, and a variety of reasons that people enter the classroom that do not have to do with a desire to politically engage with others. Yet, while feminist tinkerers must take care not to ignore the material realities that impact people's lives day to day, focusing on the process instead of on final products can allow for a kind of pedagogical play that resists a capitalist-focused end product model.

In the feminist science shops I run in some of my classes, we work with social justice organizations, allowing the class to focus on research process in a move away from traditional research goals. Decolonial scholars make clear that researchers must understand accountability as understanding the inequitable distribution of resources (through colonial theft); therefore, taking action toward redistributing material resources must be part of the research agenda (Patel 2015). That is, it is not enough to identify the inequities; researchers should be involved in taking action. This resonates with many feminist methodological approaches that see action as a key part of all research and pedagogy (Crawley, Lewis, and Mayberry 2008). In my classes, I focus on process as a way to create some real long-standing challenges to the material distribution of resources and life chances. However, I also directly make sure to fund each of my community partners in all the work I do beyond academic borders. This is a way to acknowledge the wealth accumulated in academic settings and the need for direct community control. In this moment, having access to money is hugely important for survival.

In my feminist science shop classes, we work throughout the term on a collective research project. Toward the beginning of the project, we brainstorm and come to consensus around our assumptions and guiding principles for the project. I always make sure to include in this an orientation to the project in which we are more committed to the ethical process than getting to one particular place. This opens up other questions of accountability for the students because we are working with a social justice organization on a specific question that they have asked us to research. The students (and I, too, probably) have a desire to create a useable report, grant proposal, or data for the organization that we are working with. The stakes of the work seem to take on a different feel than academic work that might have an overall slower pace for an end result (although not for frequent and rapid publication pressure).

The reminder to slow down and think more about the process fulfills another aspect of our desire for accountability. The slower, process-focused work also has resulted in new directions for research and information that we probably would not have found if we were too narrowly focused on producing a particular answer. For example, my class investigated the correlation of Latina/o/x breast cancer incidence and environmental toxins in California. A narrow focus on the end result would have left out incarcerated people, a huge portion of the population of California, because we could not find official, published, or quantitative data. Instead, the class decided to follow the path that the lack of research pointed to and slowed down to collect stories from formerly incarcerated women and report that there was a lack of data. We found that other researchers, in a rush to produce numbers and final results, make a population already marginalized become literally invisible in the case of breast cancer. So, while not a playful example in the sense of fun that we associate with play, this more winding research path took us to important political and potential scientific findings.

The focus on process does not mean we do not aim to change material conditions. Quite the opposite is true. We resist the urge to focus on a traditional kind of output or product—publishable research in this case. Instead, the focus on ethical engagement, even if it is slower and takes us in unexpected directions, is at its root committed to a real tangible goal: creating better health outcomes for the population we are working with. Changing the way students think and interact and the revaluing of what kind of knowledge is considered trusted is part of creating this real world.

A limitation in using playful methodologies in these classroom contexts is that it might seem strange to use the word "playful" to describe work focused on breast cancer or gunshot wounds. This is a place where both the transfeminist artists work and Lather's work on "getting lost" help us to think about play as a kind of anti-productivity based in searching and producing new connections between ideas and data. Further, play in the sense of creating relationships outside the direct work on research or activism has been an important aspect of breaking down some of the boundaries of classroom walls. We make sure to share a meal with our collaborators from the social justice organizations when we can. This builds new kinds of relationships with the students and collaborators as well as for me as the professor.

Love, then, is part of our knowledge production process, as María Lugones has argued. She is not talking about science proper, but she gives us a way to think about knowing that can be applied to our understanding of who we are. Lugones (1987) theorizes playfulness as a way to lovingly know other people by traveling to their worlds and a way to be open to self-reflection with the possibility of changing ourselves and the multiple worlds that we inhabit. She is careful to describe how each of the concepts she uses—specifically "worlds," "at ease," "love," and "playfulness"— have simultaneously different meanings. Lugones refuses to provide a fixed definition for "world" because she says, "the term is suggestive" and she does not "want to close off the suggestiveness of it too soon" (9). Instead she describes some possibilities for what is included in her conception of "worlds." To her, worlds can match with the dominant constructions of geographic and/or cultural boundaries, can counter these, can exist within each other, and can exist as overlapping. One way to know that you are existing in different worlds (which she argues one can do at different times or even simultaneously) is you are a different person in each world— you have different attributes or ways of being. In some worlds you are more at ease than in others. The attribute she focuses on to illustrate the existence of these worlds and the ability of individuals to "world travel" is playfulness. She finds that in some worlds she can be playful, and in other worlds she cannot.

We can take multiple lessons from Lugones' playfulness. First, these concepts— in our case, tinkering or biohacking (Malatino 2017)—should not be essentialized as inherently good or bad. Second, tinkering as a form of playfulness can be used to learn more about our worlds, ourselves, and others. This could be an orientation to the process of knowledge production (methodology). This playfulness then always has to do with how we are situated in relation to the world or "subjects" that we are learning about and learning to identify with. Lugones's playfulness offers a possible adumbration of Haraway's (1988) situated knowledges. In Haraway's formulation of situated knowledges, she argues for a feminist objectivity that is always about partial vision from a particular location. Lugones ends her essay arguing for a "disloyalty to arrogant perceivers, including the arrogant perceiver in ourselves, and to their constructions of women" (1987, 18). This argument against "arrogant perceivers" is an argument against the idea of a pure objectivity. Instead, she provides a language for understanding how "objective" truths (her interest is in truths about women) are never able to be made at a distance, in part because of the relationality and cultural world that creates us and we create. Lugones's playfulness serves as a call for creativity in the sense of playful co-creation. Her engagement with the idea of standpoint theory or situated knowledges resonates with activist claims for a special view from the location of the oppressed and marginalized together with the importance of a collective practice of knowledge production.

Lugones's "loving perception" is not about neutrally observing the world. It is a situated kind of understanding that takes one's own subjectivity as well as others' subjectivity into account. In earlier chapters, I cautioned against taking positive affective relations to science (love, passion) out of context as evidence for the need to engage with science without questioning what we mean by "science." This instead gives us a glimpse into a specific way to think about love. This kind of decolonial love has been theorized as a way toward new knowledge-making and new world-making in a broad sense. Specifically, decolonial love can offer a way toward an ethical engagement with science. Taken together with Lather's *Getting Lost*, the irrationality of love has the potential to bring us to new, unexpected results. It is a call for a kind of openness and risk-taking; at the same time, decolonial love might offer us a way forward that also has to do with healing.

Roy (2012) uses Lather's proposition to "work the ruins" (2007) of science, asking what we can do to keep moving forward with feminist science after the disappointments of science. Calling these "disappointments" may not get the point strongly enough. Science has been and continues to be a violent, colonial institution. As Roy notes, it wounds. She argues that we must engage with it knowing that this will involve "self-wounding." This leaves open the question of how to heal from such wounds. Colonial science has no doubt left large wounds, and continuing to engage with such science means we risk opening up these same wounds and creating new wounds. Leaving science on its own, however, does not mean it will not continue to wound. The challenge for us as feminist tinkerers is to engage with it in such way that transforms our knowledge systems, our worlds, and ourselves. Perhaps decolonial love can be a way to heal these wounds while acknowledging their existence and power.

Carolyn Ureña (2017) contrasts colonial and decolonial love with colonial love being "based in an imperialist, dualist logic, [that] dangerously fetishizes the beloved object and participates in the oppression and subjugation of difference" (87). Decolonial love, on the other hand, through its "acceptance of fluid identities and a redefined but shared humanity" (87), "encourages a form of ethical intersubjectivity premised on imagining a 'third way' of engaging otherness beyond Western binary thinking" (88). A decolonial love might be one that acknowledges the kind of intimacy inherent in any kind of colonial violence. Love in the colonial sense has been suggested to forget this violence and instead believe in the benevolence of colonial forces for the good of the colonized. A decolonial love is one acknowledging the harm, violence, and difference between knowledge producers. It is a love that can acknowledge what it cannot know and take responsibility for it does know. Working in the ruins may be a way to work through the intimate space of violent knowledges.

Ureña (2017) argues that, as part of a decolonial project, decolonial love has the potential to be part of "healing the psychological, affective, and epistemic wounds occasioned by the hierarchical division of the world into colonizers and colonized, a split implicit in the concept of modernity/coloniality" (88). Ureña qualifies the use of "healing": "Rather than imply a total erasure of past wounds the way that 'transformation' or 'cure' might, healing in the decolonial context relentlessly underscores the ethical dimension of this necessarily ongoing practice, a process that need not even be realized in order to remain a worthwhile venture" (89). Feminist tinkerers can apply this understanding of healing through continued challenges to colonial science's wounding practices, rather than simply trying to use colonial science in the reverse for social justice. This approach, the "third way" that comes from Sandoval's theorizing on love (2000), means that we must create something new out of an ethical and playful engagement with science.

Quimera Rosa did lament the fact that working in a U.S. context and holding workshops in a university setting seemed to limit the kind of connections that participants were making with each other and their group. They were surprised that students came to the workshop at the agreed upon time and then immediately left at the end of the session. The kind of playful work they do, which has to do with relationship building, means that they often continue tinkering with materials and/or ideas after designated times. Once again, there is a difference between a fluid relationality that is about building anticapitalist communities and ways of being and the corporatization of this dynamic to get the most work out of individuals by building unhealthy relationships where workers feel a need to work long hours and exhibit proper affective attachment to their coworkers and their company. In the neoliberal versions of tinkering where DIY biologists used their free time to tinker, these biologists invoked playfulness even as their goals remained capitalist focused. Neoliberalism has skillfully co-opted the idea of passionate play and community building in its flexible response to the 1960s and 1970s anticapitalist rebellions.

Lugones (1987) argues that love needs to "be rethought, made anew" (7). She argues that the love that commonly circulates is based on abuse and servitude. Instead, she argues for a love that is about learning to identify with other women across difference. She is specifically interested in women of color loving one another across difference. She poses it in opposition to an arrogant perception of the other while at the same time warning against losing one's own self by simply trying to identify with another. For her, love and identifying with another is about relationality. Likewise, Sandoval in her invocation of "love" as a decolonial practice argues against Western narratives of love and instead thinks of "'love' as a hermeneutic, as a set of practices and procedures that can transit all citizen-subjects, regardless of social class towards a differential mode of consciousness and its accompanying technologies of method and social movement" (2000, 140). The purpose of this form of love is the transformation of ourselves and worlds through political action. Although Sandoval argues for everyone being able to access this kind of methodology, we must continue to be vigilant about how this can slip into a kind of universalized thinking that ignores real material differences.

The use of love in the work of James Baldwin has also been claimed as a form of decolonial love (Drexler-Dreis 2015). Baldwin argues for love as a way to deal with racism, not by moving past difference but rather by confronting it along with its production through a history of slavery and continued racism. He argues for love not as a simple emotion but sees it as something more and in a political sense.

> A vast amount of energy that goes into what we call the Negro problem is produced by the white man's profound desire to not be seen as he is, and at the same time a vast amount of the white anguish is rooted in the white man's equally profound need to be seen as he is, to be released from the tyranny of his mirror. . . . Love takes off the masks that we fear we cannot live without and know we cannot live within. I use the word "love" here not merely in the personal sense but as a state of being, or a state of grace—not in the infantile American sense of being made happy but in the tough and universal sense of quest and daring and growth. (1992, 95–96)

The kind of love that Baldwin argues for is about complexity and can be read as in line with a kind of practice of traveling, playfulness, openness, and relationality. The kind of growth he talks about can be seen similarly to Lugones's as a way for transformation of self and others through the practice.

Love in these ways is about humility, learning, and vulnerability in many ways. It makes sense then that an ethical playfulness could lead to this. To really see other people, we need to understand ourselves as always interrelated with others. Therefore, we become different in these loving interactions. How does one become a feminist scientist capable of embracing shift shaping in an ethical way? Lugones (1987) mentions "the trickster and the fool are significant characters in many non-dominant or outsider cultures" (13); in her essay, they represent the ability to take on multiple selves as one travels worlds.[4] Lugones comes back to this idea of foolery in describing aspects of playfulness: "Playfulness, is in part, an openness to being a fool, which is a combination of not worrying about competence, not being self-important, not taking norms as sacred and finding ambiguity and double edges a source of wisdom and delight" (17). In this description, there are certain resonances with the dominant vision of the tinkerer, the amateur DIYer who is trying things out without care for the rules and is challenging the need for an institutionally validated competence. However, the kind of playful fool that Lugones describes is one who is humble and focused on self-reflection. These are not merely side notes but are central to a feminist embrace of tinkering.

In the introduction to Lugones's *Pilgrimages/Peregrinajes* (2003), which includes a republished version of her essay on playfulness, she explains that playfulness to her is "at the crux of liberation, both as a process and as something to achieve" (33). This highlights how tinkering can be a process where new knowledges and possibilities are produced as well as being a way of being or becoming. The humility and self-reflective nature of Lugones's playfulness is not present in the entrepreneurial, goal-oriented, dominant version of the tinkerer. The DIY community biologists interested in business incubation seemed to desire freedom to individually succeed at capitalism. Neoliberalism also shapes us into different kinds of people—those conditioned to imagine the only kind of freedom we could desire is capitalist success. Breaking out of the rules/logics of Western philosophy and science is not part of the imagination DIY biologists can access. Lugones's playfulness does not ignore material injustices, but she is ultimately concerned with the systemic injustices brought about by colonization and how patriarchal and racist practices dominate our worlds. Her playfulness is an act of rebellion against these that aims to dislodge their power and create new realities.

In my feminist science shop classroom, I am constantly challenged (as are my students) by working within an institution that is not interested in decolonial change. One of the functions of the university is to justify difference in life outcomes by acting as an arbiter of meritocracy. Societal divisions are drawn between those who are deemed to be deserving of inclusion in the university; then, once in the university, those who are more or less deserving of success are ranked through grades. Entrance to universities and the awarding of grades are supposedly fair and neutral processes, but it is well-known that this is not the case due to systemic injustices that begin before birth for our noncollege and college students.

One quarter, my class surprised me with a pushback to the system of grading by arguing for a collective grade at the end of the quarter (Giordano and Cruz 2019). I was caught off guard because, although this would benefit many students, some students may have done a lot more work than others. I would have thought that they would be upset at not getting "what they deserved." The students argued collectively that they thought this was the fairest way to do the grading because we were all working collectively during the course. This did not solve the deeper problems of inequity, but it showed me that it was possible in a short amount of time (ten weeks) for students to build a different kind of community, even within a colonial institution of the academy—itself a material result of the science shop.

The division of those worthy and unworthy of higher grades translates not only into numbers or letters on a piece of paper (a transcript) but can be the rationale for paying workers different amounts of money. Therefore, the division of students at every stage has real material consequences in our economic world (Meyerhoff 2019). Again, this does not mean that one class changed this significantly; however, it is an example of the potential of working within the ruins of the university to create new kinds of spaces.

Work from a newer thread of critical university studies at the intersection of ethnic studies points to the contradictions and possibilities in the noninnocent university system, which is best understood as a "set of relations" (Singh 2021). The process I took the students through during the ten-week course created a specific kind of relationality and an accountability to each other and the larger world through our partnership with the group California Latinas for Reproductive Justice. What may have seemed fair and just at the beginning to some no longer seemed so. The students semi-jokingly told me that they had become "socialists" while making their case for why collective grading was fair. They argued against the idea that the fairest contribution to the project was every student putting in the same number of hours. They knew they each had different life circumstances that were overdetermined by identity-based inequalities. Some students may have been working multiple jobs while other students had their schooling paid for by their parents. This was one way that the time each could contribute to the project might not be equal in number but could be fair in a broader sense of the term. This also demonstrated to me that I could not predict at the beginning of the course where we would end up or who we would be.

In a direct engagement with the question of feminist methodologies for producing scientific knowledge, Roy (2012) offers a rich feminist scientific praxis that comes back to this idea of openness, risk-taking, and playfulness through the idea of "feeling around." Roy begins her chapter by foregrounding the importance of a feminist researcher thinking "carefully about her relationship with what is to become the known" (313). She sets out to define a new way of thinking about the self-reflexive praxis of feminist science-making: "In the context of the natural sciences, issues of ontology become intertwined with issues of ethics when we realize that the 'me' in a situated knowledge has always been an 'us' and, particularly in the research laboratory, that this 'us' includes the nonhuman as well as the inorganic" (314). Lugones focused on playful world traveling between women as a way to know the "other," but Roy's articulation of the other as more than human or even living provides an entry point to apply Lugones's theorizing of playfulness to scientific praxis. Lugones's world traveling, then, makes possible Roy's desire to "position the knower in the same critical plane as that which becomes the known" (313–314).

Ethical meaning-making is central to the playfulness of feminist science practice. Roy (2012) draws heavily on Lather's (2007) feminist methodology of "getting lost" to argue that we should engage with moving into difficult territories, specifically those of scientific knowledge production, with an openness and introspection. Lather meditates on the losses of "an innocent science" as well as "the unitary, potentially fully conscious subject; researcher self-reflexivity as a 'way out' of impasses in ethics and responsibility; and transparent theories of language . . . also . . . the clear political object" (2007, 12) after poststructuralism and postmodernism. By delineating this loss of a clear path, with a clear place of beginning and a clear end point, she sets up getting lost as an ethical position of acknowledging the impossibilities of controlled experiments. In this proposal, getting lost is not about succumbing to a simple relativism and disavowal of any agency; instead, Lather and Roy bring to the front the idea of becoming responsible and accountable for what we learn to know.

What Roy adds to this openness to knowledge production is Haraway's (2008) concept of "becoming with"—to reread the well-known feminist scientist Barbara McClintock's concept about having a "feeling for the organism" (Keller 1983). Roy argues that we can take this idea in two different directions at the same time. First is the more common application to argue that a feminist scientist might learn something new by relating to her subject, by learning to feel "for" one's subjects (or objects of inquiry). However, Roy changes the inflection to ask, What if we learn to "feel around" for our subjects? She proposes that this kind of playful experimentation allows us to see the moment of knowledge production as a specific co-producing of us (the researchers), the matter under study, and the knowledge. Roy nods to Karen Barad's (2007) work on agential realism in which Barad argues that there is no pre-existing "known" to be discovered but rather the experimenter, the apparatus, and the matter in question are all produced in a specific moment. That truth during a specific "agential cut" is based on the choices made in that moment that include those of the researcher, the larger political and cultural forces at play, and the matter in question. Roy therefore gives us a language and a methodology to work toward becoming response-able in these intra-actions.

Lugones's playfulness, similarly, is about being open to changing oneself, worlds, and others. In *Pilgrimages/Peregrinajes* (2003), Lugones begins the introduction by describing a "practice of *tantear*": "The 'pilgrimage' that the title of the book calls forth moves through different levels of liberatory work in company forged through a practice of *tantear* for meaning, for the limits of possibility; putting our hands to our ears to hear better, to hear the meaning in the enclosures and openings of our praxis" (1). Lugones explains, "I use the Spanish word 'tantear' both in the sense of exploring someone's inclinations about a particular issue and in the sense of 'tantear en las oscuridad,' putting one's hands in front of oneself as one is walking in the dark, tactilely feeling one's way" (1). The first part of Lugones explanation of tantear is about a politics of partial vision or situated knowledge, a view from "somewhere" instead of "nowhere." Tantear for Lugones suggests a playful yet cautious feeling around in the dark and a searching for connections. Taken together, these uses reinforce both the idea of self-reflection and the idea of forging new connections outside of oneself. This concept also resonates with Lather's (2007) "getting lost" while feeling around in the dark as a praxis of meaning-making.

This kind of "feeling around" (Roy) or "playful world traveling" (Lugones) is about praxis. Lugones explains that she uses the term praxis to "mark the indissoluble link between theory and practice" (2003, 37). The two cannot (and should not) be separated. These concepts are often thought of in opposition or as presenting a choice between thinking and really doing. Feminists have long challenged this idea (e.g., hooks 1991). Play is active. By approaching feminist research playfully, we open our work to the power playfulness might have to create new knowledges and ways of being and interrelating, Through play, feminist tinkerers are well-positioned to recognize the contextual and partial nature of any knowledge that we can hope to find. What are we playing with, what do we make through our play, and how exactly do we enact playful research?

FEMINISTS' HANDS-ON PEDAGOGIES

Approaching hands-on learning within a feminist genealogy places the practice as not something newly discovered. However, we must acknowledge the historical moment in which this idea regains traction for a particular kind of learning and knowledge production. This moment of interest from science educators and practitioners in getting more people involved with playing with science through hands-on exploration finds its parallel in the humanities through a move to deal with matter itself. These humanities interventions, for example, new materialisms, have been characterized as a challenge/rejection to postmodernism based in part on a frustration with the failure to stop imperialist wars and corporate globalization (Povinelli 2016).

A feminist genealogy for feminist tinkering can help us to challenge a common assumption or at least slippage that follows from our current humanities/science consensus on interdisciplinary hands-on science—that is, that there is something more "real" that can be discovered through touching "matter" itself or engaging with scientific research. Instead, an approach to hands-on praxis is needed that highlights the ethics, politics and knowledge we gain from a DIY, skills-sharing *process* itself. Clare Hemmings (2012) argues that "for feminist theorists [the] question of process is a political as well as methodological concern, in that it seeks to enhance knowledge and create the conditions for transformation through an engagement with others across difference" (151). This articulation of process resonates with Lugones's playful world traveling as well as other theorists and activists who focus on the interactions between humans (and nonhumans). Process is about how we move through the world, which has political, ethical, and material consequences. A focus on process is itself a kind of materialism, but not one that is focused on dealing with matter itself in a depoliticized way that simply gives agency to it as if it is a separate entity; instead, it is a materialism that is concerned with the distribution of material resources. This is a concern with how gendered racial capitalism and colonization have taken control of resources such as land, water, and food and produced divisions that produce differential life chances for humans and have destroyed the chances of survival for nonhuman animals and our planet as a whole.

The feminist praxis of hands-on work requires testing and producing new theoretical understandings, and it is always politically situated. Examples from various activist movements demonstrate that the process of knowledge production matters. Hamraie (2017) argues that disability activists produced crip technosciences through processes of tinkering and often physical rematerialization. In the

TINKERING AS A FEMINIST PRAXIS   163

case of "access" through curb cuts, they argue that activist curb cuts—that is the curb cuts that were physically made by activists before government regulations standardized the curb cut—"propose[d] access as negotiation, rather than as a resolved, measurable end" (102). This activist intervention was a practice of tinkering, a practice in iteration.

While valuing and revaluing marginalized knowledges, how might we also use affective politics to understand feminist tinkering as a methodology that can be learned across difference while not ignoring the material realities of differing social locations?

Hemmings (2012) offers one suggestion for this through the concept of "affective solidarity." Hemmings's arguments stem from an understanding of affect as political and situated. Hemmings begins by suggesting that the idea of "empathy" as the primary affective attachment as the basis for a coalitional politics is limiting, arguing that it may act to ignore differences and their roots by assuming a universalizing idea of womanhood that is interested and best served by empathetic relations across difference. Following others, she argues that this may strengthen these categories instead of disrupting them. Instead, Hemmings poses the idea of affective dissonance as a shared precursor to coalition building through affective solidarity. Hemmings points out, "Affective dissonance cannot guarantee feminist politicisation or even a resistant mode. And yet, it just might . . . that sense of dissonance might become a sense of injustice and then a desire to rectify that. Affect might flood one's being and change not only how the house and its circumstances are experienced and understood, but how everything else is seen and understood too, from this time on" (2012, 157). Therefore, Hemmings argues that while feminist politicization is not the usual result of affective dissonance, affective dissonance is necessary for feminist politicization. This politicization that stems from a knowing (or maybe a feeling, or a knowing-feeling) that something is "not right" can be the basis for a solidarity between feminists.

Gloria Anzaldúa's concept of *la facultad* can help to situate Hemmings's notion of affective dissonance at the intersections of feminist standpoint theories, highlighting that this dissonance is an ongoing process or capacity rather than a single moment that breaks everything, "*La facultad* is the capacity to see in the surface phenomena the meaning of deeper realities, to see the deep structure below the surface ([1987] 2012, 60). Anzaldúa argues that this capacity to see reality is more likely to develop in those who experience oppression. She argues that it is out of necessity to protect oneself by anticipating the harm that might happen to them; it is a "survival tactic" (61). Anzaldúa writes, "[*la facultad*] is latent in all of us" (61). Her description of how this capacity develops resonates with the idea of affective dissonance.

> Fear develops the proximity sense aspect of *la facultad*. But there is a deeper sensing that is another aspect of this faculty. It is anything that breaks into one's everyday mode of perception, that causes a break in one's defenses and resistance, anything that takes one from one's habitual grounding, causes the depths to open

up, causes a shift in perception. This shift in perception deepens the way we see concrete objects and people; the senses become so acute and piercing that we can see through things, view events in depth, a piercing that reaches the underworld (the realm of the soul). (61)

This development occurs from a dissonance between the way the world and we, ourselves, are supposed to be and how we sense it or see it. She continues, "We lose something in this mode of initiation, something is taken from us: our innocence, our unknowing ways, our safe and easy ignorance" (61).

La facultad and affective dissonance circle back again to the idea of knowing differently from below and arguments for feminist standpoint theory. However, both ideas also work beyond the idea of the rational/irrational borders on which colonial, Western thinking is founded. This combination of knowledge production that includes affect is not a simple argument to take emotion and experience as replacements for scientific epistemologies. Sensing, feeling, and knowing are not separable in these constructions. Instead, these ways of engaging affect offer a "third way" as Chela Sandoval (2000) calls for in her use of "love."

In practice, political tinkering is often more about the process than a specific end result. It is a common part of feminist pedagogy to center praxis and addressing real-world problems (Crawley, Lewis, and Mayberry 2008). In all my feminist science studies classes, I introduce some direct interaction with scientific publications. In one graduate course where I create a feminist science shop through the class, we partner with a social justice organization to take up Weasel's (2001) call to incorporate this work directly into the classroom. My work with the People's Community Medics (PCM) is one such example. PCM is an organization that was founded after a community-based investigation into the 2009 murder of Oscar Grant found that ambulances were not responding in a timely manner to life-threatening events in predominately Black Oakland neighborhoods. In response, PCM was founded to train community members to be first responders in cases such as gunshot wounds because time to treatment has been shown to be crucial for survival. This idea was based in self-determination and communities caring for themselves, which we can see echoed from earlier Black liberation and civil rights health movements. The research we undertook in collaboration with them was to find out more about how people in their specific community were surviving gunshot wounds so that they could incorporate that information into their trainings and activism and learn from those who have survived (Cruz et al. 2019).

My students attended the organization's training and became first responders ourselves, learning hands-on the techniques and working directly with community members on this question as well as learning more about the medical field from our literature reviews and conversations with doctors, nurses, and first responders. This praxis opened up questions for us about how to produce new knowledges ethically and responsibly.

I stress in my classes how the process itself ends up being the most important part of the project. Through this, we not only produce different knowledges that could be missed but also become different ourselves and thereby are shaping new relationships and worlds. The goal of this is not to become more realized individual subjects; quite the opposite, the relationality that is learned has the potential to challenge who students see as part of their communities and therefore who they have political, ethical, and, importantly, economic responsibilities to.

In community DIY workshops the practice of gathering and building political community is often as important or more important than producing a specific end product. For example, Christina Dunbar-Hester's (2014a) ethnography of pirate radio activists in the 1990s–2000s, revealed that, although there were disagreements at times among activists, for many the political goals of the project made the end result of having usable equipment less important than the political community building that was able to take place during technical skill sharing.

I found a similar result when I participated in a feminist soldering workshop run by Dr. Ellen Foster (2016). During the workshop, we aimed to build contact microphones—which we did do. However, what I took from the workshop was the conversations we had while tinkering together. Sitting among the instructors and other participants, we had time to discuss our relationships with soldering, and science and technology in general. During these discussions we learned individually and collectively how our relationships with tinkering revealed specifically gendered, raced, and classed relationships. The process of sitting down together and tinkering was not about getting to any place in particular. Feminist tinkering can be about the process itself.

A similar result was produced in the DIY classes I took at mainstream community science laboratories. However, these groups did not specify a political vision, thereby differing from the activist communities. What we learned in the space was not necessarily useful for creating new scientific knowledge; what was produced was a deeper affective attachment to science and specifically DIY science. Without a basis in social justice politics, more access to science (imagined as objective and innocent) becomes a political and ethical goal in and of itself.

For Quimera Rosa, who are self-identified feminist biohackers, hands-on artistic practice opened up more ways to think about the relationships between their own bodies, others, and the experimental materials, animals, and plants. In this way they followed the kind of feminist science methodology proposed by Roy (2004) that includes considering one's relationship with one's subjects and objects of study. Quimera Rosa's project, *Transplant*, began as a way to experiment with the relationship between humans and plants. The impetus was a concern about how humans continue to be centered and imagined as distinct entities in environmental debates. The artistic, experimental project of *Transplant* is to become a human-plant hybrid by injecting chlorophyll intravenously. The outcome is not to create a distinct new kind of life but rather to learn from the questions that arise as they move through the scientific protocol to accomplish this task. Questions arise constantly throughout the process, from the beginning of thinking about the project, to picking plants to extract chlorophyll from, to adding the possible medical uses of chlorophyll for treating cancer, to discomforts around animal research, to tensions arising about the inseparability of humans and plants, to trying to do science (or art) outside of capitalism. Quimera Rosa's *Transplant* project also pushes the boundaries of hand-on experimentation by conducting self-experimentation—using themselves instead of animals as the subjects.

In another kind of transfeminist project, oncogrrrls performances are used as ways to answer questions about breast cancer. The performances take place in various locations around the globe, and in each case the questions are collectively derived and then collectively answered through a collaborative performance. Through the physical and participatory process, the audience individually and collectively learn new things (Novella 2017, 2022).

To highlight the ways that the process may produce something other than a physical scientific discovery is not to suggest that these feminist participatory experiments do not have the potential for producing scientific results or new truths more broadly. The hands-on and collective experimentation characterized by each of these feminist science examples demonstrates possibilities for engaging with science in ways that remake how we understand science, our bodies, our collectives, and our worlds. Hands-on work for feminist tinkerers is about world-making and remaking. By acknowledging the hands-on working with material that takes place as part of the experiment, we allow a different way of potentially understanding and becoming response-able for our results.

Feminists' Iterative Experimentation

In mathematics, to iterate is to repeat an operation. Part of the process of tinkering is iteratively checking one's work and trying slightly new approaches. This is not necessarily a progressive process where the result gets better or where you can accurately plan the next iteration for a specific desired result. Instead, each iteration produces a new configuration, engaging in a process of assembly, assemblage, and reassembly. Murphy has used these concepts to think about the entanglement of the feminist women's health movement in science and medicine and neoliberal governmentality. This tinkering with what exists can also be seen in Lather's (2007) methodology of "working from the ruins" that Roy (2012) takes up.

Roy begins from scientifically published material and reanalyzes it, tinkers with the title, and creates a new engagement with the data, thereby reassembling it and producing new knowledge, theorizing that "perhaps the key is to keep on using the scientific method with the tools and technoscientific practices that she has at hand until a 'subversive repetition' of the scientific method takes her to somewhere new" (2012, 12). Quimera Rosa similarly discussed using scientifically published protocols but reworking them based on their own ethical interpretation. The Black Panther Party reassembled the ideas and people who participated in the government's "war on poverty" programs into new formations of democratic control for Black communities.

Reassembly is not about working from the outside but about the outcome of a certain grouping of people, ideas, and other materials in a given moment. Disability activists, for example, tinkered with already built environments to produce new formations to understand the concept of activist iteration and reassembly as not only an ongoing process but one that illuminates tensions. Hamraie (2017) argues that the relationship of disability activists to tinkering and redesigning technologies represented "a technological ambivalence" that in "itself [was] a disabled way of knowing-making, born from the iteration of lived experience, technological failure, and ambivalence toward the fantasy of normalization" (107). The space for iterative reassembly opens up new possibilities, including for conversations about the tensions that come up with new formations. These acknowledgments and conversations represent a democratic, participatory approach.

Performance and performativity can each produce new realities through iteration. Quimera Rosa brought up each of these concepts as central to their work during our interview. They suggested that the events that they referred to as performances instead of experiments were still about creating new knowledge. They argued that the performance of the experimental protocol—going through the steps—opened up new ways of thinking and being. Interestingly, their group began not as a biohacking group but instead their main interest in hacking had to do with gender and sexuality. Their first work was engaged in genderhacking and genderf-cking in which they drew their ideas from queer theorists such as Judith Butler.

Butler's intervention into understanding gender and sex through performativity gave us a way to think beyond the nature/culture debates. Instead of understanding gender/sex as binary, innate categories or as split into a cultural component (gender) and physical/natural component (sex), Butler (2006) proposes that we physically manifest and become gendered through our repetition of acts. This opened up space for the genderhacking and genderf-cking that Quimera Rosa took up by iteratively repeating new acts and seeing what kinds of gender we might create. It is based in these ideas of performativity and genderhacking that their work moved to exploring genderhacking beyond human forms, such as *Transplant*.

Caro Novella describes oncogrrrls's performances as *rehearsals* not as in preparation for a "final performance" but "within the experience of bodily attuning to the not-yet-known, new sensoriums, attentional practices and viscous possibilities developed with.in" (2021, 13). Her praxis suggests the importance of not knowing, unknowing, and repetition in creating new knowledges. Who is in the room for each rehearsal changes the questions and answers they collaboratively produce about cancer.

Rehearsals of oncogrrrls allow for consideration of ethics throughout each iteration, what could be called a practice of feminist diffraction. Although the practice of feminist reflexivity has been useful in many cases, including feminist pedagogical practices, feminist science studies scholars have critiqued the early adoption of reflexive practices in science studies for the ways that the knower and known remained separate and how reflexivity often relied on the idea of reflection with an analogy that can restabilize the idea of being able to look back at an accurate image of the self and the past. Haraway (1994) argued for the limits of reflection and reflexivity in favor of diffraction instead for providing patterns of difference instead of sameness and mirroring. Barad expanded on this metaphor in pointing to the limits of reflection and arguing for a diffractive methodology: "The point is not simply to put the observer or knower back *in* the world (as if the world were a container and we needed merely to acknowledge our situatedness in it) but to understand and take account of the fact that we too are part of the world's differential becoming. And furthermore, the point is not merely that knowledge practices have material consequences but that *practices of knowing are specific material engagements that participate in (re)configuring the world*" (2007, 91, emphases in original).

Diffraction makes visible the differences that matter in different intra-actions (Barad 2007). The concept of intra-action versus simply action moves us away from the idea of individuated agency and instead toward ideas of entanglements. Through practices of diffraction, we understand each experiment as specific to the conditions in which it takes place and that the observer and the observed are co-produced through the experiment. As Barad explicated, knowledge production itself is part of the "(re)configuring" of the world. This resonates with other conceptual tools proposed by feminist science studies scholars such as Haraway's (2008) *naturecultures*, which highlights that the world is always co-produced and that there is an inseparability of nature and culture, and more specific ones such as Angie Willey's *biopossibilities*, which gives us a way to think about "species- and context-specific capacity[ies] to embody socially meaningful traits and desires" (2016a, 555). These feminist science studies concepts offer us ways to understand each experiment as specific and culturally and politically situated. This kind of diffractive methodology is at play (pun intended) in Quimera Rosa's projects, in which they iterate on scientific protocols that already exist to call attention to the choices and questions that come up in each instance.

Ethics figures prominently in feminist science studies.[5] Relatedly, then, responsibility and accountability are central. We are always responsible in Barad's construction of accountability: "Responsibility is not a calculation to be performed. It is a relation always already integral to the world's ongoing intra-active becoming and not-becoming. It is an iterative (re)opening up to, an enabling of responsiveness. Not through the realisation of some existing possibility, but through the iterative reworking of im/possibility, an ongoing rupturing, a cross-cutting of topological reconfiguring of the space of response-ability" (2010, 265). The idea of ability for response is about "how" we intra-act; it highlights the way that humans (as well as nonhumans) produce specific entanglements. The process of becoming response-able is not about examining something we have done from a distance but rather understanding ourselves as a part of the results. To understand knowledge production in this way (refusing to split object and subject) might be a way, as Haraway says, "to become answerable for what we learn how to see" (1988, 583).

Iteration in feminist tinkering is about response-ability: responsibility, ethics, and accountability are integrated in the process of knowledge production so that the results themselves are partially determined by the way we approach the practice of knowing. For Lugones (1987), playful world traveling is a way to create coalition by understanding others and ourselves through these interactions (or intra-actions?) in different worlds. Similarly, Haraway's idea of response-ability is about a relationality to others (human and nonhuman) and an understanding that we become—and become undone—in relation.[6] If, as Lugones argues, we are different beings in different worlds, then playfulness is about being open to new becomings with the understanding that who we are and who others are is never static and always emerges from our intra-actions.

Conclusion

How might we understand the political and ethical investments of feminist science tinkerers? How might we move forward in creating these intentional political and ethical communities or coalitions? While collectives and communities of activist scientists are formed and have formed, identities and groupings are also always in flux, always iterative. One form that iterative experimentation can take is feminist tinkering: a way to ethically engage through concepts of playfulness, world traveling, decolonial love, affective solidarity, feeling around, and getting lost. Feminist tinkering offers an opportunity to challenge fixed biologies and political formations by both acknowledging their power and iteratively making new ways of worlds and beings.

Epilogue

If we were not sure before, the COVID-19 pandemic uncovered the inability of Western Science and Democracy to keep us alive. In theory, the nations with the supposedly most advanced, most well-financed science industries would be best positioned to deal with a novel biological virus. Science did not save us. But then again, that might depend on who we mean by "us"?

Most of this book was researched and written before the COVID-19 pandemic. It is about the politics and practices of democratic sciences in the mid-teens of the twenty-first century. So what has happened after the COVID-19 pandemic at the various sites of democratic science that I analyze throughout this book—in DIY biology community labs, feminist studies classrooms, and activist/art spaces?

Although claims were made that the virus "does not discriminate," the COVID-19 pandemic laid bare the structural inequalities that many know first-hand. Ruth Gilmore's definition of racism as "the state-sanctioned and/or extrale-gal production and exploitation of group-differentiated vulnerability to premature death" (2007, 28) was painfully proven accurate. In the United States, it was hard to ignore this systemic racism while a disproportionate number of people of color labored under dangerous conditions and a disproportionate number of white people stayed home, protecting themselves and their families.

The impact of the gendered and racial divisions of labor became clearer. The care labor or reproductive labor that had been relegated to the feminine could not be passed on to the lowest paid workers because of safety concerns regarding the well-being of the employer's families. Housekeepers, maids, and nannies were "let go." Teachers were suddenly responsible for long-distance babysitting of the nation's kids at the same time as taking care of their own homes and families. Women across racial groups were both laid off and resigned more often than men (Chotiner 2020). This was presumably because of the continued gendered divisions of labor that forced women to do more care labor at home and also assumed that men's jobs were more important to protect because they needed to continue to provide finan-cial support for the family.

EPILOGUE

The idea of American democracy as a "government of the people, by the people, and for the people" rang hollow as decisions were made based on capitalist profit, resulting in huge gains for the already rich in a year that left many newly unhoused, unemployed, and without adequate food. Worldwide, the promise of globalization as good for humanity was likewise unequivocally shown to result in differential life chances between so-called developed and developing nations when the distribution of life-saving vaccines were controlled by multinational corporations that financially profited while millions died preventable deaths.

Millions literally took to the streets to protest the racist killings of Black people during the pandemic. Although the direct murder of Black people at the hands of the U.S. police was the immediate target of protests, many made connections to the larger issues of premature death that included COVID-19 deaths. At the same time, people began to help each other through thousands of projects that made connections with the politics of mutual aid based on various intersecting theories of Indigenous ways of living/surviving, anarchism, and direct democracy. Mutual aid collectives came into being to provide for the most basic, urgent needs of communities with the long-term anticapitalist goals of transforming society so that our needs can always be met through cooperation (Spade 2020). For example, collectives began distributing food to community members who needed to stay home because of underlying health conditions or who were unemployed and could not afford food.

The pandemic provided for a devastatingly honest reckoning of the underlying structures of our society and the possibilities for new worlds built on collective survival. Gendered racial capitalism was more clearly exposed, and through that exposure alternative worlds became more possible.

I did not return to my research sites directly after the pandemic, but I kept up with the labs, academics, and activist communities from afar, and I witnessed new possibilities open. The DIY biology movement took action through ways that sometimes intersected with activist mutual aid projects and the call to reckon with anti-Black racism. For example, at Genspace a six-week course on racism and science has been offered twice since 2020 as well as a 2020 feminist reading group on cyborgs and speculative fiction. Perhaps this opened space for the questions I raised in the final chapters about which genealogies democratic sciences draw on. In another example, one of the founders of the Open Insulin Project explained that it exists at the intersection of both mutual aid and the more recent DIY biology movements (Talbot 2020). Other DIY biologists jumped into action to try to make testing and ventilators more accessible, and there was hope that many of their goals to make science more transparent would come true.

Not only did we see the impact in specific DIY and activist spaces, broader conversations about science and the role of the public became ubiquitous. The participation in COVID-19 debates was great (as in magnitude, not necessarily in substance). Everyone seemed to become an expert on COVID-19, its transmission, and how to protect against it. In some ways, this moment showed the potential for a truly democratic, participatory science literacy. It was hard to find someone who did not know something about COVID-19 and did not have any opinions about it. Whether one feared viral infection or not, COVID-19 showed that viruses could not be purely biological—the pandemic was clearly an inseparable assemblage of the biological, political, and economic. Science could not simply be done in laboratories and presented to the public. Perhaps this moment showed not only the potential for a liberatory interdisciplinary science literacy but the desperate need for it.

We also saw how important affect was in determining what people believed and how groups formed with similar beliefs. While fear and distrust of Science dominated emotions on one side, the antidote offered to this was love and trust of Science. This demonstrates that continued binary ways of thinking dominated. In contrast to this dichotomy, critical science movements calling attention to both the limitations of science and technology, the cultural embeddedness of these fields, and the potential for promoting (human and nonhuman) life-affirming politics through science and technology have always existed, and in this moment might be more needed than ever. Even before the pandemic, there was a reboot of Science for the People chapters along with its magazine.[1]

One of my hopes for this book is that it can help us make sense of this complicated landscape. I have made the case that we must be critical of the use of both social justice rhetoric (e.g., March for Science protests) and impulses to declare one's love for science (e.g., I F-cking Love Science website and Facebook page) when they do not challenge systemic inequalities because embracing such moves may actually act to strengthen those inequitable structures and protect them from critique. Instead, I have offered feminist tinkering as a form of critical science literacy. Here, science literacy does not mean the mainstream form of increasing appreciation for science. Critical science literacy to me is a way to "rewrite" scientific knowledge (Wynter 1984), thereby taking a much more active role in knowledge production.

Critical science literacy has been associated with a radical politics of "love and rage" (Weinstein 2010). While I wholeheartedly agree with the importance of Matthew Weinstein's mode of critical science literacy aimed at claiming activist writings as knowledge-making projects, I have taken a different angle in defining and practicing critical science literacy.

Previously, I published case studies of my own undergraduate and graduate teaching in which students in the academy are exposed to critical studies and basic science literacy to gain the ability to both critique existing scientific results and produce their own new knowledges (Giordano 2017b). I added the feminist science shop model to this kind of critical science literacy to break academic/nonacademic borders and connect my students with local social justice activists as they produce activist community–driven research (Cruz et al. 2019). These methods have been somewhat successful on a small scale as I have been able to exploit the university's current interest in community engagement and interdisciplinary research.

As this book has cautioned, uncritical use of universal pedagogical interventions makes this model vulnerable to quick co-optation if we try to scale up too quickly. We must remember that the university is neither innocent nor static. The

EPILOGUE

university is deeply implicated in settler colonialism, slavery, gendered racial capitalism, and other oppressive systems often imagined to be separate from it. At the same time, many have used, hacked, and exploited the university toward our own disruptive aims (Harney and Moten 2013; la paperson 2017). This is where tinkering comes in for me. The lessons I drew on in chapter 6 from feminist, decolonial, and queer theorists about using multiple approaches and constant iteration will serve us well in these efforts.

I have been asked by others how we teach critical science literacy or how we set up a science shop. I am hesitant to make a list of steps or a how-to guide because of the need for constant iteration. Instead, there are two principles I will offer for those of us who attempt to do this work from the academy. First, one must be involved in social justice movements—on the ground, in the street, and probably these days on the internet. In the academy it is hard to maintain these real connections because of time, elitism, and (related to each of those) the requirements of our job for promotion.

Part of adhering to this principle may require moving away from the orientation that makes us successful in the academy, that of self-aggrandizement and focus on individual success. Instead, we need to be able to embrace a collective orientation. Questions that may arise and should among academics (particularly those with job security in tenured or tenure-track positions) are about one's investments in academic success, fame, and even the economic benefits of promotion. Whether or not we commit to activist collectivism, our choices no doubt impact a collective.

Questions about our own choices should come with the question, At whose expense do I do this work and for whose success? Being part of a collective invested in social change, instead of seeing oneself as the intellectual figuring out the cleverest new way to analyze our world, is not simple. After all, we have to write single-authored manuscripts such as this one to maintain our positions. I offer these principles as ones I struggle with and aspire to.

Although I have argued that to succeed in this first principle—of true collectivity with radical social justice organizations outside of the academy—we need to let go of some of our individualism, the second principle is about using one's unique, individual privileged position. This second principle I offer is to work to redistribute the material resources that we have access to in the academy. I do not offer this as a call for heroism, which would go against the purposes of enacting these principles. Instead, this is a call to understand one's agency and limits in determining responsibility to tinker with the system.

This principle of using one's privileged position has to do with understanding what resources are at your disposal and trying to redistribute them. As I mentioned in chapter 4, I came to feminist studies through privilege and oppression politics. I have developed my political and academic understanding of these politics in the last twenty years, but I find some of the more basic ideas useful to go back to for taking action. In this school of thought, those who have access to unearned privileges—for example, through their jobs—use those positions to help redistribute resources. For example, if you have a job with access to a copier, you make copies for flyers and newsletters for your organizing group and offer to print things for comrades who do not have access to this.

We can take this beyond making copies in the academy to think about what we do with our professional development money and our access to library accounts, offices, and on-campus meeting rooms. More specifically for those of us attached to science through our training or science studies fields, we can think about what benefits we might have access to because of our proximity to Science. Studying S/science comes with the possibility of getting more funds from both internal and external grants. If we can secure some of these monies, I suggest we think about using this money to support on-the-ground social justice organizations and colleagues who do not have opportunities to be grant funded. In a previous position, my department decided to take the indirect funds that had come to our department because of some of my grants and make them available to all my colleagues so that the less-funded areas of research could be supported.

Throughout this text, I have analyzed who "we" is in different iterations of democratic science. Following these two principles may help those of us in academia keep a sense of belonging to our larger feminist, antiracist, social justice communities. The first principle suggests we think about ourselves outside of our professional roles in order to be part of a larger collective; the second principle then pushes us to ask how our professional benefits can aid the larger collective that we understand ourselves as belonging to. These principles also must operate on multiple levels: they are about working to enact local changes while understanding global contexts and conditions for possibility.

And, of course, I come back to the importance of tinkering and iteration. This work must be a constantly changing process in communication with communities inside and outside academia. Quite simply, my argument is if we wish to strive for more democratic sciences, we cannot possibly do it without a strong commitment to redistribution of epistemic and economic privileges, benefits, and resources.

Acknowledgments

I am grateful for all the folks who supported me along the way in moving this book from a proposal that began in 2015, through many phases of tinkering and rewriting, to a published work today. Over this time, I moved across three different academic positions and cities to make my way back to my political, queer community in Atlanta, Georgia. There were so many humans who contributed to this book over this near decade journey that I must apologize in advance to any folks who I unintentionally leave out of these acknowledgments.

First, a huge thank-you to my colleagues, including those at San Diego State University, University of California at Davis, and Kennesaw State University for talking through ideas and finding funding to support this work at various stages. I developed many of these ideas through talks at conferences, invited guest lectures, and job talks—even those where I was not extended final offers. To those who did the labor to make these events happen, who invited me, attended, and asked difficult questions, thank you. I benefitted greatly from the financial, intellectual, and emotional support from my collaborative experiment with Rana Jaleel and Tim Choy, "HATCH: Feminist Art and Science Shop," a project funded by the University of California–Davis Mellon Research Initiative. Furthermore, thank you to the undergraduate and graduate students who helped with data collection and analysis over the years, in particular Yi-Lin "Eli" Chung and Ilana Turner.

Second, I owe thanks to the activist-scholars/scholar-activists who have paved the way for those of us who continue to hold on to this boundary-crossing work with all of its complexities, complicities, and possibilities. Also, I appreciate the DIY biologists, biohackers, transfeminists, activists, organizers, artists, scientists, students, and academics with whom I engage in this book and the experiments they have put out into this world for all of us to play with, learn from, critique, reflect on, iterate, and analyze. My attention to how your work comes into this world, proliferates, transforms, mutates, and offers possibilities for new kinds of worlds is meant to be generative. Special thanks to those who allowed me to interview you and spend time in your lab spaces.

Additionally, many people labored to read parts of this work over the years and provide feedback. A sincere thank-you to Nadia Behizadeh, Rajani Bhatia, Liz Constable, Gwen D'Arcangelis, Steve Giordano, Rebecca Herzig, Lilly Irani, Rana Jaleel, Clare Jen, Ingrid Lagos, Kerrie Lynn, Sarah McCullough, Andrea Miller, Caro Novella, Deboleena Roy, Banu Subramaniam, Kalindi Vora, and Angie Willey. Special thanks to those who read the entire manuscript and those who read chapters over multiple iterations. Also thank you to the anonymous manuscript reviewers—I do not know who you are, but I could not have arrived at this point without your insightful suggestions.

Furthermore, I am grateful to the professionals in the publishing world who guided this book to production in the last few months. Thank you to Kim Guinta and the editorial staff, in particular Emma-Li Downer, at Rutgers University Press for all of your support, guidance, and labor in seeing this book to publication. Thank you also to the editors at the University of Washington Press for the work you did to improve this manuscript over the years we worked together, although we ultimately did not see it to final publication together. I do not want to obscure the ways that a final product often leaves out the challenging path it took. All the ups and downs are part of this final product. Thank you to Matteo Farinella for creating the cover art: I appreciate not only the beautiful final piece but the patience, wisdom, creativity, and care that you shared with me through the design process. Also, thank you to the editors at Flatpage and Ideas on Fire for your labor through your professional eyes to create a better book.

Finally, I am grateful for both the intentional and unintentional support that my friends, families, kids, lovers, comrades, and nonhuman companions have provided over the many years of working on this project. Although he was not able to see this book take root, I appreciate my Grandpa Isi for planting a seed when I was still a child by saying I would make a good researcher, to whatever possibilities that meant and may mean in the future.

Financial support for part of this work was provided through the National Science Foundation (grant number 1456707). Parts of this book iterate on arguments that I began in previous published work: "New Democratic Sciences, Ethics, and Proper Publics," *Science, Technology, and Human Values* 43 (3) (2018): 401–430; "Those Who Can't, Teach: Critical Science Literacy as a Queer Science of Failure," *Catalyst: Feminism, Theory, Technoscience* 3 (1) (2017): 1–21; and "Theorizing Feminist Tinkering with Science Methodologies," *Canadian Journal of Science, Mathematics and Technology Education* 18 (2018): 222–231.

Notes

PRELUDE

1. For a recording of Senator Clinton's response to Dan Rather on *CBS Evening News*, September 13, 2001, see BoogieWithStew, "Hillary Clinton: Every Nation Must Be with Us or against Us," YouTube, October 24, 2007, 0:15, https://www.youtube.com/watch?v=DbYGY iGjpUs. The text of President George W. Bush's speech is found in "Address to a Joint Session of Congress and the American People," The White House [archive], September 20, 2001, https://georgewbush-whitehouse.archives.gov/news/releases/2001/09/20010920-8.html.

INTRODUCTION

1. By "in the streets" I am referring to protest activity, as in "take to the streets" or "reclaim the streets." (For instance, see Le Tigre's 2001 song "Get Off the Internet" for the popular lyrics "I'll meet you in the street; Get off the internet!" LeTigreWorld, "Get Off the Internet," YouTube, July 19, 2018, 3:32, https://www.youtube.com/watch?v=YHf28w7LB8c.) I am not referring to living on the streets as in being unhoused, or learning by the streets or on the streets, or being "street smart" as in knowing a kind of useful common sense that comes from living in a so-called rough neighborhood (as in poor and racialized as non-white).

2. In chapters 1 and 2 I more fully define synthetic biology and address the question of whether synthetic biology represents a new field or is simply genetic engineering rebranded.

3. The original image was created by Kristin Joiner. It can still be found along with a brief history behind the sign at Embolden Wisconsin, https://www.emboldenwi.org/kindnessiseverything.

4. Many feel that the sign represents a kind of virtue signaling instead of a true investment in engaging in political movements. In an op-ed, Amanda Hess (2021) argues that the sign is liberal white women's response to critiques of white women in the face of a Trump victory. Interestingly, related to the DIY topic of this book, the sign originated as one woman's handmade sign that was made and placed outside after Trump's election. You can now buy the sign from multiple vendors as well as many alternative signs with the same aesthetic.

5. Science has continued to be a site where knowledge about who are legitimate political and economic subjects are in part produced and reproduced, with some variations over time. For example, scientific work continues to be used both for and against ideas of sex/gender essentialism, arguing that either men and women are two separate, distinct natural

177

categories or science shows more variation than simple binary sex/gender. What is at stake in these debates are often political rights as well as recognition through biology of being "normal." In some cases, these fights are for physical existence itself: for example, the struggle to stop surgeries on intersex children who do not fit into binary categories of female or male. Battles over scientific truths therefore take place both inside and outside the traditional scientific settings of the university and more recently Big Bio (pharmaceutical and other biotech corporations).

6. For example, in the United States when slavery was challenged in the nineteenth century, resulting in a period of Reconstruction, a violent backlash based on racism derailed democratic, cross-racial challenges to capitalism (Du Bois 2017). Capitalism further responded with great flexibility in the first half of the twentieth century by providing more of a safety net to some workers while leaving many Black workers out—slightly shifting yet maintaining racial lines by assimilating Eastern and Southern European immigrants into the category of Whiteness (Brodkin 1998; Faber 2020; Katznelson 2006). Some European nations fought back against anarchist and socialist/communist threats with fascism or state-controlled communism based on ideas of national belonging and staying in line for the good of the state (e.g., Germany, Italy, Russia, and Spain). Therefore, racial/national lines were reinforced, and racially "desirable" women were largely relegated to providing reproductive labor for the good of the nation. In the United States, another key strategy was institutionalizing unions and thereby keeping tighter reins on worker organizations.

7. *Oxford English Dictionary*, s.v. "tinker (v.)," September 2023, https://doi.org/10.1093/OED/6083867409; *Oxford English Dictionary*, s.v. "tinkerer (n.)," July 2023, https://doi.org/10.1093/OED/7961848403.

8. For example, I responded (Giordano 2018) to a recent call for a special issue of the *Canadian Journal of Science, Mathematics, and Technology Education* focused on embracing tinkering for social justice, contrasting engineering and tinkering as a way to challenge traditional scientific objectivity.

9. Claude Lévi-Strauss (1966) introduced the concept of the *bricoleur* versus the engineer to describe the idea of working with materials at hand and repurposing for different means, as opposed to creating and using tools specifically designed to solve an engineer's problem. In the last half-century his idea has been both taken up widely and criticized widely. Most notably, Jacques Derrida argued that neither the bricoleur nor the engineer could be innocent subjects and that, once the myth of Western objectivity is removed, the engineer is also a bricoleur (Spivak 1976). I do not directly follow the path of *bricolage* and *bricoleur* although they have often been translated as "tinkering" and "tinkerer." No doubt the tinkerer of today is influenced by the bricoleur of other disciplines. I am more focused on the current uses of tinkering in biology. Many features of today's tinkering biologists will likely resonate with those who study the bricoleur's movements in the humanities over the last decades.

INTERLUDE 1 — SERENDIPITY

1. My dad recently told me that what I did not remember from that story was that Serendipity Association, Ltd., had also found a way to get public services—specifically, the Department of Sanitation—to clean up for them as they were building. My dad told me about how he took pictures of NYC sanitation workers cleaning the area and called the department to report it. Before he could show the pictures, he was told that he was mistaken: what he saw was normal, appropriate business taking place a few blocks away. Once he sent them the pictures, they had to admit that they were cleaning up for a private business and subsequently they charged Serendipity.

NOTES TO PAGES 23–28

2. It is unclear from my research whether the fires took place before or after the property purchases. I found 167 records in the NYC Property Database ACRIS for Serendipity Association, Ltd. These records included the purchase and later sale of properties all within about a ten-block area surrounding the block in question in the 1980s and 1990s. It was more difficult to find the fire records. Therefore, the fires could have been set after the purchase of homes in an area that was already significantly economically gutted through a familiar neighborhood killer: the building of a freeway on Third Avenue. Although the roadway was elevated, it still caused a dark separation between the blocks below Third Avenue to the waterfront with its quickly dying industries and those blocks above Third Avenue. The idea of waterfront revitalization was popular among gentrifying organizations. This makes the idea of buying property cheaply below Third Avenue while working to revitalize the area and then developing said properties a potentially profitable idea. There are many cases of landlords and insurance agents collaborating in New York City in the 1970s and 1980s to collect large insurance settlements on properties damaged from arson. Therefore, I include this information to leave open the possibility that the fires were serendipitous not for the ability to purchase these properties cheaply but perhaps to profit from inflated insurance plans and then have the properties leveled in order to develop in the future.

3. Also, I want to acknowledge that this tale of arson and profit is conjecture based on the recollection of neighborhood activists who were aware of what was happening in the area. There is good reason to believe these stories, however; I will point out that, regardless of the veracity of who set the fires and why, people were losing their homes, the neighborhood was in a depressed state, and the idea of finding something "lucky" (serendipitous) in your purchasing power of vast amounts of property at low cost is disturbing—at least to me.

4. *Oxford English Dictionary*, s.v. "serendipity (n.)," July 2023, https://doi.org/10.1093/OED/1563393800.

5. Skid Row is a neighborhood in Los Angeles well-known for its large population of unhoused people.

6. One example of an activist intervention has been the occupying of homes by a group called Moms4Housing in Oakland, California (https://moms4housing.org/). Other activist organizations such as Gay Shame have actively resisted gentrification in San Francisco through various means.

CHAPTER 1 — (DE)CONSTRUCTING DIY COMMUNITY BIOLOGY LABS

1. The thirteen labs were BioArt Laboratories, BioCurious, Biologigaragen, Biospace, BOSSLab, Brightwork, BUGSS, DIY MadLab, Genspace, LA Biohackers, La Paillasse, Open Wetlab, and Symbiotica. Some of these did not have accessible websites when I conducted the analysis. There are examples of text and images from the websites throughout this chapter. I also included Pink Army Cooperative and BioBricks in the original analysis as two prominent synthetic biology projects that were not community DIY lab spaces but were democratic synthetic biology projects. I later included analysis from the DIYbio website and archived email mailing list. DIYbio is not a lab but rather an organization to bring together DIY projects across the world.

2. This number grew to 111 by May 2021 although it is unclear how active the website and mailing list are now. Worldwide the list includes twenty-six in the United States–East, sixteen in the United States–West, eight in Canada, forty-one in Europe, nine in Asia, six in Latin America, and five in Oceania, with notably none identified in Africa.

3. Text from the original 2012 "Our Vision" page on the BUGSS: A Place for Creative Biology site (https://web.archive.org/web/20130331025751/http://www.bugssonline.org/our-vision.html). This organization is presently active at https://bugssonline.org/.

180 NOTES TO PAGES 29-58

4. All Genspace quotations can be found on the 2013–2016 version of the website's "About" page (https://web.archive.org/web/20130912185110/http://genspace.org/page/About). Genspace is presently active at https://www.genspace.org/.

5. Respectively, from the original (2013–2015) "Our Mission" page of LA Biohackers (https://web.archive.org/web/20130813065043/http://biohackers.la/about), and the 2014 iteration of the "About BOSSLAB: Boston's Open-Source Science Center" page (https://web.archive.org/web/20140812095056/http://bosslab.org/about-bosslab/). The LA Biohackers site is now defunct; BOSSLab, presently known as BOSLab, is located at https://www.boslab.org/.

6. From the 2012–2015 version of the BioCurious "About" page (https://web.archive.org/web/20120127033334/http://biocurious.org/about/) as well as the blog site at https://biocuriosity.wordpress.com/about/. BioCurious is presently found at https://biocurious.org/, and the blog continues to be updated at https://biocuriosity.wordpress.com/.

7. As seen in the entry for La Paillasse on the Hackerspaces wiki (https://wiki.hackerspaces.org/La_Paillasse), last updated July 1, 2022. The organization's website is no longer functioning.

8. Quotation from the main page of Biospace from 2012 to 2014 (https://web.archive.org/web/20120322032744/http://www.biospace.ca/); this website is presently defunct. Quotation from the "About" page of BioArt Laboratories at http://bioartlab.com accessed and recorded on June 14, 2014. This website is presently defunct.

9. TAXA Biotechnologies, "Glowing Plant Kickstarter video," July 24, 2013, 2:07, https://www.youtube.com/watch?v=YxFQ9MkwbDs.

10. Genspace was in the building known as the Metropolitan Exchange Bank building although it was never a bank. The owner had a good amount of press and notoriety for trying to keep somewhat affordable loft space for startups and artists in a real estate market that was becoming more and more unaffordable (e.g., see Lipinski 2011).

11. See his website that has links to much of his writing and narratives about his own life, https://www.karymullis.com/.

12. VitruvianMan07, "iGEM - Drew Endy Defining Synthetic Biology (video)," YouTube, June 13, 2007, 5:00, https://www.youtube.com/watch?v=XIuh7KDRzLk.

13. Similar debates have taken place in the larger hacker/maker communities over the decision to use U.S. military funding through DARPA (Defense Advanced Research Projects Agency) to put on the Maker Faire. Although at least one notable hacker, Mitch Altman (2012), vocally shared concerns and quit association with the event over this decision, the movement as a whole continues with DARPA funding.

14. There has been quite a lot of concern and debate about the role of genetic ancestry testing and its now public role in racial identity-making. For example, see Nelson (2016) and TallBear (2013).

CHAPTER 2 — THE TINKERER AS A NEW SCIENTIFIC SUBJECT

1. Stephen Jay Gould's popular book on this topic is *The Mismeasure of Man* (1996).

2. Hacker, maker, and tinkerer are all used in the self-identification of community biology lab members. My use of tinkerer instead of hacker or maker to characterize this subject-figure formation is chosen not because it is necessarily the most common term for self-identification of individual members of the groups I examine. Rather, I chose it to argue that this term, which is prevalent in these communities, is useful in understanding the ties to earlier forms of tinkering and experimentalism and to politically situate it. That is, I am not interested in who calls themselves tinkerers or not and why, but instead I trace how this idea of modern-day DIY biologists "tinkering" produces a specific subject-figure that I call *the tinkerer*.

NOTES TO PAGES 60–76

3. For example, Lawrence Summers felt confident in defending not only a belief in natural aptitude but in arguing that it indeed followed a gendered pattern and explained disparities (Barres 2006).

4. Steve Jobs was a co-founder of Apple Computers, the first company to sell a personal home computer. The popular story of his success is that he began the company in his parent's garage in his twenties after dropping out of college. He eventually became a multibillionaire.

5. I name heterosexual pairings specifically because there is evidence that there is a greater degree of cross class intimacy and family built through gay communities than in heterosexual communities (Stacey 2004).

6. From the 2013–2016 version of Genspace's "About" page (https://web.archive.org /web/20130912185110/http://genspace.org/page/About).

7. "Spare time tinkering" is a quote from Ellen Jorgensen, co-founder of Genspace (Kean 2011).

8. Not all Google employees are white males, but as of 2015 their increased diversity efforts had resulted in a workforce that was only 31 percent female, 2 percent Black, 3 percent Hispanic, 32 percent Asian, and 59 percent white (https://www.blog.google /topics/diversity/focusing-on-diversity30/). Shifting demographics of tech industry jobs may show the flexibility of racial capitalism and require a more nuanced understanding of racism and capitalism in the current moment. See Aihwa Ong's *Neoliberalism as Exception* (2006) for one such account.

9. Future research tracing genetic laboratory work through the night is needed to verify this hypothesis.

10. This is not to say that universities and their embrace of inclusion politics have solved deep inequities. The diversification of the university through this approach has often been through tokenizing moves that bring certain members of marginalized groups into positions of power in ways that they cannot challenge the status quo; the politics of inclusion are limiting. Still, diversity reforms in the sciences have actually been long fought for battles that I have argued elsewhere represent possibilities for deeper changes to epistemological diversity (Giordano 2014).

INTERLUDE 3 — LEARNING THE LIMITS OF ETHICAL DEBATE

1. Stephen M. Siviy, Nicole J. Love, Brian M. DeCicco, Sara B. Giordano, and Tara L. Seifert, "The Relative Playfulness of Juvenile Lewis and Fischer-344 Rats," *Physiology and Behavior* 80, no. 2–3 (2003): 385–394, https://doi.org/10.1016/j.physbeh.2003.09.002.

CHAPTER 3 — BECOMING THE INFORMED PUBLIC

1. Emphasis on the original page. O'Reilly: BioCoder, ca. 2013–2018, accessed June 2024, https://www.oreilly.com/biocoder/.

2. Quotations from "About" on the BioCurious blog site (https://biocuriosity.wordpress .com/about/) and "FAQ: Safety" on BioCurious.org and the blog site (https://biocurious .org/faq/#safety and https://biocuriosity.wordpress.com/faq/).

3. "Codes," DIYbio, ca. 2011, https://diybio.org/codes/. This page hosts the draft codes of ethics from the European (May 2011) and North American (July 2011) congresses.

4. Considering the U.S. military's actions and scope, "hostile" may be a better descriptor for the United States in world politics but the United States and other Western nation-states continue to create lists of threats and "terrorist" organizations.

5. "Our Name," LA Biohackers website, ca. 2013–2015 (e.g., https://web.archive.org /web/20140803152841/http://biohackers.la/name); the LA Biohackers is now defunct.

6. The LA Biohackers "Habitat" page can be viewed at https://web.archive.org/web /20140803171541/http://biohackers.la/laboratory. The photos, dating February 28, 2013– January 24, 2014, including "NOT A METHLAB," are found on the Flickr photostream of user Cory Tobin at https://www.flickr.com/photos/58051182@N02/.

7. "Local," DIYbio, ca. 2024, https://diybio.org/local/.

8. "Open Wetlab Workshops," Waag FutureLab, 2024, https://waag.org/en/open-wetlab -workshops/; "About," Biologigaragen, ca. 2014 https://web.archive.org/web/20140425075045 /http://biologigaragen.org/about/ (the Biologigaragen website is now defunct); "About," Genspace, ca. 2013–2016, https://web.archive.org/web/20130912185110/http://genspace.org /page/About.

9. Emphases mine. Statement on the homepage of Genspace, ca. 2010–2017 (e.g., https:// web.archive.org/web/20101114015155/http://genspace.org/); "Our Vision," BUGSS, ca. 2013– 2016 (e.g., https://web.archive.org/web/20130401040507/http://www.bugssonline.org/our -vision.html).

10. From "How Can I Get Involved?" on the "Frequently Asked Questions" page, BUGSS, ca. 2015–2017, https://web.archive.org/web/20150227204123/http://www.bugssonline.org/faq .html.

11. "Giving," UCDavis, College of Biological Sciences, ca. 2014–2016, https://web.archive .org/web/20160927000519/http://biology.ucdavis.edu/giving/index.html.

12. Quotation from the main page of Biospace, ca. 2012–2014, https://web.archive.org /web/20120322032744/http://www.biospace.ca/ (this website is now defunct).

13. From "An Introduction: BioArt Laboratories" on the 2013–2014 versions of the home page (e.g., https://web.archive.org/web/20141217025259/http://bioartlab.com/).

14. From, respectively, the BioCuriosity blog home page (https://biocuriosity.wordpress .com/) and the 2014 version of the Biologigaragen "Participate" page (https://web.archive .org/web/20140425073924/http://biologigaragen.org/participate/).

15. For example, out of 691 synthetic biology labs identified by the Synthetic Biology Project's map inventory there were thirteen community labs versus 327 university-based labs.

16. Some of the early DIY synthetic biologists began DIYbio (https://diybio.org) as a way to keep the DIY community loosely connected. I use "DIY biology" to reference the larger movement and not the organization DIYbio specifically.

17. "Codes," DIYbio, ca. 2011.

18. I use Science with a capital "S" here to evoke the idea of Western science as an institution granted with the highest epistemic authority, implicitly assumed to be the best truth telling system. Harding (1991) distinguishes between the idea of multiple sciences (small "s") and the capital "S" I have mentioned. Poststructuralist scholars also often use the capital letter to signal that one variation of said institution or idea is dominant without being specified—that is, to signal and thereby challenge the embedded assumption of its universalizeable, natural dominance.

19. Based on self-disclosure through https://diybio.org, there is one identified in Bangladesh on one page of the website and a couple in Indonesia on a different page of the website.

20. From "Adult Education and Outreach" on the "About" page, Genspace, ca. 2013–2016.

21. From the BioCurious blog home page at https://biocuriosity.wordpress.com/.

22. From "What Resources Does BUGSS Have?" on the "Frequently Asked Questions" page, BUGSS, ca. 2015 (https://web.archive.org/web/20150227204123/http://www.bugssonline .org/faq.html), and the 2014 iteration of the "About BOSSLAB: Boston's Open-Source Science Center" page (https://web.archive.org/web/20140812095056/http://bosslab.org/about-bosslab/).

NOTES TO PAGES 92–115

23. "History" and "Mission" on the original Biotech without Borders home page (ca. 2018) at https://web.archive.org/web/20180330105054/http://www.biotechwithoutborders.org/.

24. Statement on the home page of BioBricks Foundation, ca. 2012–2016 (e.g., https://web .archive.org/web/20120503070441/https://biobricks.org/): "We believe fundamental scientific knowledge belongs to all of us and must be freely available for ethical, open innovation. *This is a new paradigm*" (emphasis on site).

CHAPTER 4 — FEMINIST LABS OF OUR OWN IN ACADEMIA?

1. I am exploring the usefulness of defection here as a metaphor, but this is not how any of the specific individuals I claim as defectors identify. I am using it to examine our relationships to science and feminism as feminists who are formally trained as scientists and to think through the importance of how we describe this relationship.

2. Perhaps due to the use of the term with regards to nation-states and therefore citizenship, Ruha Benjamin (2016) introduces the term "biodefectors" to describe resistance to the concept of biological citizenship, which is most used in regards to the use of genetic technologies.

3. I say this based on the amount of federally funded grants available for science research compared to nonscience research as well as the startup packages that faculty in the sciences receive compared with other fields. Even with fears and some cuts to federal science budgets under the Trump administration, the salaries and funding from graduate student stipends to faculty startup packages have continued to demonstrate this split.

4. The formation of interdisciplines such as ethnic studies and feminist studies arguably both opened up new epistemic possibilities and reconsolidated old epistemic power through incorporation (Ferguson 2012).

5. It is also important to not ignore the fact that feminist studies has been an overwhelmingly white field in the academy. In fact, part of my argument is that the way we have approached feminist science studies re-creates racialized power through a knowledge production that continues to hold scientific authority on a pedestal.

6. I use *real* in quotation marks here to draw attention to what I see as a problematic distinction between thinking about discursive analysis or critique of science as less important and less real than doing some kind of hands-on science project. Although this exercise may lead to new forms of knowledge, there seems to be some unexamined assumptions about how physically touching matter reveals truth. These assumptions may actually be based on reliance back to the ways that science itself is given ultimate epistemic authority in describing our bodies and worlds. That is, matter is assumed again to give us unmediated access to Truth (Willey 2016b).

7. The way that interest in the "natural world" is conflated with Science should be noted here again. Presumably, when interest in the natural world is assumed to be shown by interest in science classes, other activities such as gardening are likely not considered to be part of knowledge-producing interaction with the natural world. Or would arts that incorporate views of the "natural world" be considered science? These seem to be practices that have to do with some kind of love of "nature" but may not translate to enthusiasm in science courses.

8. This is not a purist stance against anything that science and technology have produced. It is quite difficult in our current world system to imagine never using modern medicine or technologies such as phones or computers. The kind of resistance that I am highlighting is in specific instances, where possible, to offer counternarratives and to ask questions instead of automatically assuming that scientific explanations are the most useful for their understanding of the world.

184   NOTES TO PAGES 119–146

INTERLUDE 5 — WHEN THE RIGHT COMES TO THE DEFENSE OF SCIENCE

1. Airaksinen, Toni. 2017. "Feminist Prof Says 'Traditional Science' Is Rooted in Racism," *Campus Reform*, October 24, 2017, https://www.campusreform.org/article/feminist-prof -says-traditional-science-rooted-racism/10021.

CHAPTER 5 — TOWARD QUEER SCIENCES OF FAILURE

1. Walter Mignolo (2015) characterizes Wynter's call for a new science through the term "decolonial scientia" to also decouple from the tradition of Science. Mignolo's character-ization of *decolonial scientia* (118) intersects with my interest in a feminist science as he argues that it attends to the role of history in creating our physical and biological realities at the same time as working toward creating new knowledges through what I would call a situated knowledges (Haraway 1988) approach.

2. The video is UCT Scientist, "Science Must Fall?" October 13, 2016, 4:14, https://www .youtube.com/watch?v=C9SiRNibD14.

3. Deboleena Roy (2004) uses the term "feminist spy" in passing to describe the chal-lenge of being a feminist pursuing her PhD in the natural sciences.

4. Also much of the work done by scientists in proper science spaces is not conducted in laboratories either. Lead researchers spend much of their time writing grants, journal articles, and supervising student workers and laboratory assistants rather than working directly with hands-on experimentation.

5. The etymology of "laboratory" traces back to the medieval Latin *laboratorium*, which was a place to *labor*. It is unclear to me why scientific workspaces became rather exclu-sively called laboratories for the last few hundred years until recently other collaborative knowledge production spaces began appropriating the term. Other places of laboring did not seem to take up the term laboratory. And eventually laboring became associated with professions requiring physical labor while intellectual work became distinct and less asso-ciated with the term labor. So, while laboratory and labor are related through etymology, the labor politics of science laboratories today deserve special attention here for the ways that so much invisible labor takes place within and beyond the lab to be credited to profes-sional scientists.

6. Or in nonfeminist circles, the longing for simpler times also includes other changes to white, middle-class family structures, such as women going to work instead of staying at home.

7. Emphasis by Hong. In Lorde's published version it says "any part of those who I define as my people?" (139). I note here that Lorde's speech was originally given in 1982, demon-strating that although Keller may have characterized the mid-1980s as a time when "we" was celebrated and not as difficult to define, Lorde and other feminists of color and lesbi-ans were already critiquing easy definitions of "community." Also, Lorde had given a key-note speech to the National Women's Studies Association in 1981 pointing out the racism and homophobia that pervaded the organization and feminist circles at that time.

CHAPTER 6 — TINKERING AS A FEMINIST PRAXIS

1. Although coalitional work is being done through these networks in other places around the world, the specific phrasing of "reproductive justice" seems to be centered in the United States and Canada. However, reproductive justice movements are connected with movements in the Global South and elsewhere.

2. Rosemarie Garland-Thomson (2005) introduces the term "sitpoint theory" to point out the assumptions of a certain kind of able-bodied subject in the language of standpoint

NOTES TO PAGES 151–172

theory. Sitpoint theory as a feminist standpoint theory is not simply the view from a wheelchair but points to the importance of language in feminist theory.

3. Gayatri Spivak's essay "Responsibility" (1994) deals with concepts that Barad and Haraway build on here. Unfortunately, neither Barad nor Haraway appear to cite Spivak's work, perhaps providing an example of why my project of bringing feminist thought considered to be outside science studies proper together with those within our field might be fruitful and part of the ethical responsibility each deals with in their approach to s/Sciences. Spivak's work engages Derrida and the deconstructive turn considering the im/possibility of individual, complete, or determinable responsibility:

> I can formalize responsibility in the following way: It is that all action is undertaken in response to a call (or something that seems to us to resemble a call) that cannot be grasped as such. Response here involves not only "respond to," as in "give an answer to," but also the related situations of "answering to," as in being responsible for a name (this brings up the question of the relationship between being responsible for/to ourselves and for/to others); of being answerable for, all of which Derrida presents within the play, in French, between *répondre a* and *répondre de*. It is also, when it is possible for the other to be face-to-face, the task and lesson of attending to her response so that it can draw forth one's own. (22)

Other work has placed these thinkers in conversation around responsibility in science literacy (e.g., see Higgins and Tolbert [2018] and multiple other work by Higgins on "responseability" in science education). Further, Grace Hong (2015a) argues for the centrality of concepts of the impossible and difference through women of color feminist scholarship.

4. Haraway (1988) suggests embracing the "Southwest native American" figure of the Coyote or Trickster as a way of disrupting the idea of the world/earth as inactive and as knowledge-making as a linear and stable process.

5. Justice has recently become more important in conversations about science inside and out of science and technology studies (Mamo and Fishman 2013). Although some have pushed for justice as a more politically efficacious goal focused on the public good versus ethics as more of an individual pursuit, there is clearly no purity in either (Reardon 2013). How each is defined and employed is important to understand to make meaning of the choice of concept in any context. Here, I consider feminist approaches to ethics because of its prominence over time in literature and its contested institutional role in the sciences. I consider ethics as both an orientation and process of relationality as I describe in the main body of the text. I use justice only with modifiers such as *social, reproductive*, and *environmental*, which have gained activist meanings over time.

6. Marc Higgins and Sara Tolbert (2018) offer us practical suggestions for implementing the kind of response-ability I discuss here through a science literacy curriculum situated in decolonial/indigenous studies.

EPILOGUE

1. Science for the People was originally organized from 1969 to 1989 as a magazine and community of activists dedicated to antiwar, anti-oppression politics.

References

Adelman, Larry. 2003. *Race: The Power of an Illusion*. Video. Produced and distributed by California Newsreel in association with the Independent Television Service.

Ahmed, Sara. 2004. "Affective Economies." *Social Text* 22 (2): 117–139. https://doi.org/10.1215/01642472-22-2_79-117.

———. 2010. *The Promise of Happiness*. Durham, NC: Duke University Press.

———. 2014. *Willful Subjects*. Durham, NC: Duke University Press.

Allen, Caitlyn. 2001. "What Do You Do Over There Anyway? Tales of an Academic Dual Citizen." In *Feminist Science Studies: A New Generation*, edited by Maralee Mayberry, Banu Subramaniam, and Lisa H. Weasel, 22–29. New York: Routledge.

Altman, M. 2012. "Hacking at the Crossroad: US Military Funding of Hackerspaces." *Journal of Peer Production* 2 (July): 1–4. http://peerproduction.net/issues/issue-2/invited-comments/hacking-at-the-crossroad/.

Anft, Michael. 2017. "A Lab of Her Own: How Colleges Are Retaining Female Undergraduates in Engineering and Computer Science." *Chronicle of Higher Education*, January 22, 2017. https://www.chronicle.com/article/a-lab-of-her-own/.

Anzaldúa, Gloria. (1987) 2012. *Borderlands/La Frontera: The New Mestiza*. 4th ed. San Francisco: Aunt Lute.

Åsberg, Cecilia, and Lynda Birke. 2010. "Biology Is a Feminist Issue: Interview with Lynda Birke." *European Journal of Women's Studies* 17 (4): 413–423. https://doi.org/10.1177/1350506810377696.

Atanasoski, Neda, and Kalindi Vora. 2015. "Surrogate Humanity: Posthuman Networks and the (Racialized) Obsolescence of Labor." *Catalyst: Feminism, Theory, Technoscience* 1, no. 1 (Fall): 1–40. https://doi.org/10.28968/cftt.v1i1.28809.

Baker, Al, J. David Goodman, and Benjamin Mueller. 2015. "Beyond the Chokehold: The Path to Eric Garner's Death." *New York Times*, June 13, 2015. https://www.nytimes.com/2015/06/14/nyregion/eric-garner-police-chokehold-staten-island.html.

Baker, Pamela, Bonnie Shulman, and Elizabeth H. Tobin. 2001. "Difficult Crossings: Stories from Building Two-Way Streets." In *Feminist Science Studies: A New Generation*, edited by Maralee Mayberry, Banu Subramaniam, and Lisa H. Weasel, 22–29. New York: Routledge.

Baldwin, James. 1992. *The Fire Next Time*. Reissue edition. New York: Vintage.

Barad, Karen. 2000. "Reconceiving Scientific Literacy as Agential Literacy: Or, Learning How to Intra-Act Responsibly within the World." In *Doing Science + Culture*, edited by Roddey Reid and Sharon Traweek, 221–258. New York: Routledge.

—. 2001. "Scientific Literacy → Agential Literacy = (Learning + Doing) Science Responsibly." In *Feminist Science Studies: A New Generation*, edited by Maralee Mayberry, Banu Subramaniam, and Lisa H. Weasel, 226–247. New York: Routledge.

—. 2007. *Meeting the Universe Halfway: Quantum Physics and the Entanglement of Matter and Meaning*. Durham, NC: Duke University Press.

—. 2010. "Quantum Entanglements and Hauntological Relations of Inheritance: Dis/Continuities, Spacetime Enfoldings, and Justice-to-Come." *Derrida Today* 3 (2): 240–268. https://doi.org/10.3366/E1754850010000813.

Barr, Jean, and Lynda Birke. 1994. "Women, Science, and Adult Education: Toward a Feminist Curriculum?" *Women's Studies International Forum* 17 (5): 473–483. https://doi.org/10.1016/0277-5395(94)00040-9.

—. 1998. *Common Science? Women, Science, and Knowledge*. Bloomington: Indiana University Press.

Barres, Ben A. 2006. "Does Gender Matter?" *Nature* 442 (7099): 133–136. https://doi.org/10.1038/442133a.

Bassett, Deborah R. 2012. "Notions of Identity, Society, and Rhetoric in a Speech Code of Science among Scientists and Engineers Working in Nanotechnology." *Science Communication* 34 (1): 115–159. https://doi.org/10.1177/1075547011417891.

Bauer, Martin W. 2009. "The Evolution of Public Understanding of Science—Discourse and Comparative Evidence." *Science, Technology and Society* 14 (2): 221–240. https://doi.org/10.1177/097172180901400202.

Benjamin, Ruha. 2013. *People's Science: Bodies and Rights on the Stem Cell Frontier*. Stanford, CA: Stanford University Press.

—. 2014. "Race for Cures: Rethinking the Racial Logics of 'Trust' in Biomedicine." *Sociology Compass* 8 (6): 755–769. https://doi.org/10.1111/soc4.12167.

—. 2016. "Informed Refusal: Toward a Justice-Based Bioethics." *Science, Technology, and Human Values* 41 (6): 967–990. https://doi.org/10.1177/0162243916656059.

Bernstein, Robin. 2011. *Racial Innocence: Performing American Childhood and Race from Slavery to Civil Rights*. New York: NYU Press.

Bernstein Sycamore, Mattilda. 2004. "There's More to Life Than Platinum: Challenging the Tyranny of Sweatshop-produced Rainbow Flags and Participatory Patriarchy." In *That's Revolting! Queer Strategies for Resisting Assimilation*, edited by Mattilda Bernstein Sycamore, 1–10. Brooklyn, NY: Soft Skull Press.

Bero, Ekaterina. 2019. "Scientist-journalist Interaction: Investigating the Relationship Between the Concept of Objectivity and the Constructed Discourses on Scientific Knowledge, Particularly on Arctic Issues." MS thesis. Norwegian University of Life Sciences, Ås.

Biochemical Society. 2017. "DIY Biology—Biohacking, Citizen Science, and Community Labs." Factsheet. London: Biochemical Society. https://www.biochemistry.org/media/embdmtwg/1diy-biology-a5-booklet.pdf.

Birch, Kean, and David Tyfield. 2013. "Theorizing the Bioeconomy: Biovalue, Biocapital, Bioeconomics or . . . What?" *Science, Technology, and Human Values* 38 (3): 299–327. https://doi.org/10.1177/0162243912442398.

Birke, Lynda I. A. 1991. "Adult Education and the Public Understanding of Science." *Journal of Further and Higher Education* 15 (2): 15–23. https://doi.org/10.1080/0309877910150202.

Bleier, Ruth, ed. 1986. *Feminist Approaches to Science*. New York: Pergamon.

Bogner, Alexander. 2012. "The Paradox of Participation Experiments." *Science, Technology, and Human Values* 37 (5): 506–527. https://doi.org/10.1177/0162243911430398.

REFERENCES

Boston Women's Health Book Collective. 2011. *Our Bodies, Ourselves*. 40th anniversary edition. New York: Simon & Schuster/Touchstone. First edition published in 1970.

Brault, Claire. 2017. "Feminist Imaginations in a Heated Climate: Parody, Idiocy, and Climatological Possibilities." *Catalyst: Feminism, Theory, Technoscience* 3 (2): 1–33. https://doi.org/10.28968/cftt.v3i2.28847.

Britzman, Deborah P. 1995. "Is There a Queer Pedagogy? Or, Stop Reading Straight." *Educational Theory* 45 (2): 151–165. https://doi.org/10.1111/j.1741-5446.1995.00151.x.

Brodkin, Karen. 1998. *How Jews Became White Folks and What That Says about Race in America*. New Brunswick, NJ: Rutgers University Press.

Brown, Wendy. 2015. *Undoing the Demos: Neoliberalism's Stealth Revolution*. Brooklyn, NY: Zone Books.

Bullard, Robert D. 1993. *Confronting Environmental Racism: Voices from the Grassroots*. Boston: South End.

Butler, Judith. 2004. *Precarious Life: The Powers of Mourning and Violence*. London: Verso.

Calvert, Jane. 2013. "Engineering Biology and Society: Reflections on Synthetic Biology." *Science, Technology and Society* 18 (3): 405–420. https://doi.org/10.1177/0971721813498501.

Calvert, Jane, and Paul Martin. 2009. "The Role of Social Scientists in Synthetic Biology." *EMBO Reports* 10 (3): 201–204. https://doi.org/10.1038/embor.2009.15.

Campbell, Nancy D. 2009. "Reconstructing Science and Technology Studies: Views from Feminist Standpoint Theory." *Frontiers: A Journal of Women Studies* 30 (1): 1–29. https://doi.org/10.1353/fro.0.0033.

Caporael, Linnda R., E. Gabriella Panichkul, and Dennis R. Harris. 1993. "Tinkering with Gender." In *Research in Philosophy and Technology*, Vol. 13: *Technology and Feminism*, edited by J. Rothschild, 73–99. Greenwich, CT: JAI Press.

Chan, Ngai Keung, and Chi Kwok. 2021. "Guerilla Capitalism and the Platform Economy: Governing Uber in China, Taiwan, and Hong Kong." *Information, Communication and Society* 24 (6): 780–796. https://doi.org/10.1080/1369118X.2021.1909096.

Chotiner, Isaac. 2020. "Why the Pandemic Is Forcing Women out of the Workforce." *New Yorker*, October 23, 2020. https://www.newyorker.com/news/q-and-a/why-the-pandemic-is-forcing-women-out-of-the-workforce.

Clare, Eli. 2017. *Brilliant Imperfection: Grappling with Cure*. Durham, NC: Duke University Press.

Clarke, Jan. 2001. "From Biologist to Sociologist: Blurred Boundaries and Shared Practices." In *Feminist Science Studies: A New Generation*, edited by Maralee Mayberry, Banu Subramaniam, and Lisa H. Weasel, 35–41. New York: Routledge.

Collins, H. M., and Robert Evans. 2002. "The Third Wave of Science Studies: Studies of Expertise and Experience." *Social Studies of Science* 32 (2): 235–296. https://doi.org/10.1177/0306312702032002003.

Collins, Patricia Hill. 1999. "Moving beyond Gender: Intersectionality and Scientific Knowledge." In *Revisioning Gender*, edited by Myra Marx Ferree, Judith Lorber, and Beth B. Hess, 261–284. Thousand Oaks, CA: Sage.

Conner, Clifford D. 2009. *A People's History of Science: Miners, Midwives, and Low Mechanicks*. New York: Nation Books.

Conrad, Ryan. 2014. *Against Equality: Queer Revolution, Not Mere Inclusion*. Oakland, CA: AK Press.

Crawley, Sara L., Jennifer E. Lewis, and Maralee Mayberry. 2008. "Introduction—Feminist Pedagogies in Action: Teaching beyond Disciplines." *Feminist Teacher* 19 (1): 1–12. https://doi.org/10.1353/ftr.0.0021.

Cruz, Maya, Julia Jordan, Stacey Anne Baterina Salinas, Rebecca Jones, Sharena Thomas, Alyssa Ney, and Sara Giordano. 2019. "Using the Feminist Science Shop Model for Social Justice: A Case Study in Challenging the Nexus of Racist Policing and Medical Neglect." *Women's Studies* 48 (3): 283–308. https://doi.org/10.1080/00497878.2019.1593840.

Daston, Lorraine, and Peter Galison. 2007. *Objectivity*. New York: Zone Books.

Davis, Kathy. 2007. *The Making of Our Bodies, Ourselves: How Feminism Travels across Borders*. Durham, NC: Duke University Press.

Dean, Jodi. 2010. *Blog Theory: Feedback and Capture in the Circuits of Drive*. Cambridge: Polity.

Deleuze, Gilles, and Felix Guattari. 1987. *A Thousand Plateaus: Capitalism and Schizophrenia*. Translated by Brian Massumi. Minneapolis: University of Minnesota Press.

Delfanti, Alessandro. 2011. "Hacking Genomes. The Ethics of Open and Rebel Biology." *International Review of Information Ethics* 15 (September): 52–57. https://doi.org/10.29173/irie223.

De Rond, Mark, Adrian Moorhouse, and Matt Rogan. 2011. "Make Serendipity Work for You." *Harvard Business Review*, February 25, 2011. https://hbr.org/2011/02/make-serendipity-work.

Douglas, Heather E. 2009. *Science, Policy, and the Value-Free Ideal*. Pittsburgh: University of Pittsburgh Press.

Dragojlovic, Nicolas, and Edna Einsiedel. 2013. "Playing God or Just Unnatural? Religious Beliefs and Approval of Synthetic Biology." *Public Understanding of Science* 22 (7): 869–885. https://doi.org/10.1177/0963662512445011.

Drexler-Dreis, Joseph. 2015. "James Baldwin's Decolonial Love as Religious Orientation." *Journal of Africana Religions* 3 (3): 251–278. https://muse.jhu.edu/article/585806.

Du Bois, W.E.B. 2017. *Black Reconstruction in America: Toward a History of the Part Which Black Folk Played in the Attempt to Reconstruct Democracy in America, 1860–1880*. New York: Routledge. First published in 1935.

Duggan, Lisa. 2003. *The Twilight of Equality? Neoliberalism, Cultural Politics, and the Attack on Democracy*. Boston: Beacon.

Dunbar-Hester, Christina. 2014. *Low Power to the People: Pirates, Protest, and Politics in FM Radio Activism*. Cambridge, MA: MIT Press.

Duncan, David Ewing. 2010. "A Mission to Sequence the Genomes of 100,000 People." *New York Times*, June 7, 2010. https://www.nytimes.com/2010/06/08/science/08church.html.

Eggleson, Kathleen. 2014. "Transatlantic Divergences in Citizen Science Ethics—Comparative Analysis of the DIYbio Code of Ethics Drafts of 2011." *NanoEthics* 8 (2): 187–192. https://doi.org/10.1007/s11569-014-0197-7.

Eglash, Ron. 2002. "Race, Sex, and Nerds: From Black Geeks to Asian American Hipsters." *Social Text* 20 (2): 49–64. https://muse.jhu.edu/article/31927.

Eisen, Arlene. 2014. "Operation Ghetto Storm: 2012 Annual Report on the Extrajudicial Killings of 313 Black People by Police, Security Guards and Vigilantes." Malcolm X Grassroots Committee. Updated November 2014. http://www.operationghettostorm.org/uploads/1/9/1/1/19110795/new_all_14_11_04.pdf.

Epstein, Steven. 1996. *Impure Science: AIDS, Activism, and the Politics of Knowledge*. Berkeley: University of California Press.

———. 2008. *Inclusion: The Politics of Difference in Medical Research*. Chicago: University of Chicago Press.

Espineira, Karine, and Marie-Hélène/Sam Bourcier. 2016. "Transfeminism: Something Else, Somewhere Else." *Transgender Studies Quarterly* 3 (1–2): 84–94. https://doi.org/10.1215/23289252-3334247.

REFERENCES

Eubanks, Virginia. 2009. "Double-Bound: Putting the Power Back into Participatory Research." *Frontiers: A Journal of Women Studies* 30 (1): 107–137. https://doi.org/10.1353/fro.0.0023.

Faber, Jacob W. 2020. "We Built This: Consequences of New Deal Era Intervention in America's Racial Geography." *American Sociological Review* 85 (5): 739–775. https://doi.org/10.1177/0003122420948464.

Fabricant, Michael, and Stephen Brier. 2016. *Austerity Blues: Fighting for the Soul of Public Higher Education.* Baltimore, MD: Johns Hopkins University Press.

Fausto-Sterling, Anne. 1992. "Building Two-Way Streets: The Case of Feminism and Science." *NWSA Journal* 4 (3): 336–349. http://www.jstor.org/stable/4316219.

———. 2000. *Sexing the Body: Gender Politics and the Construction of Sexuality.* New York: Basic Books.

Federation of Feminist Women's Health Centers. 1991. *A New View of a Woman's Body: A Fully Illustrated Guide.* New York: Simon & Schuster/Touchstone.

Federici, Silvia. 2004. *Caliban and the Witch.* New York: Autonomedia.

Ferguson, Roderick A. 2012. *The Reorder of Things: The University and Its Pedagogies of Minority Difference.* Minneapolis: University of Minnesota Press.

Filomeno, Felipe A. 2017. "Gentrification Threatens Diversity of Baltimore Neighborhoods." *Baltimore Sun (Online),* June 14, 2017. http://www.baltimoresun.com/news/opinion/oped/bs-ed-op-0614-gentrification-diversity-threat-20170612-story.html.

Foege, Alec. 2013. *The Tinkerers: The Amateurs, DIYers, and Inventors Who Make America Great.* New York: Basic Books.

Foster, Ellen Kathleen. 2016. "Critical Workshopping, Sound, and the Practice of Soldering." Workshop presented at the Gender, Bodies, and Technology: (In)Visible Futures conference, Roanoke, VA, April 2016.

———. 2017. "Making Cultures: Politics of Inclusion, Accessibility, and Empowerment at the Margins of the Maker Movement." PhD diss., Rensselaer Polytechnic Institute.

Foucault, Michel. 2003. *"Society Must Be Defended": Lectures at the Collège de France, 1975–76.* Translated by David Macey. New York: Picador.

Fraser, Nancy. 1990. "Rethinking the Public Sphere: A Contribution to the Critique of Actually Existing Democracy." *Social Text,* no. 25/26 (January): 56–80. https://doi.org/10.2307/466240.

Free Radicals. 2017. "Sciencewashing the Neighborhood." *Free Radicals* (blog). November 16, 2017. https://freerads.org/2017/11/16/science-washing-the-neighborhood/.

Fullwiley, Duana. 2007. "Race and Genetics: Attempts to Define the Relationship." *BioSocieties* 2 (2): 221–237. https://doi.org/10.1017/S1745855207005625.

Garland-Thomson, Rosemarie. 2005. "Feminist Disability Studies." *Signs: Journal of Women in Culture and Society* 30 (2): 1557–1587. https://doi.org/10.1086/423352.

Geere, Duncan. 2013. "Kickstarter Bans Project Creators from Giving Away Genetically-Modified Organisms." *The Verge,* August 2, 2013. https://www.theverge.com/2013/8/2/4583562/kickstarter-bans-project-creators-from-giving-GMO-rewards.

Gentile, Dan. 2021. "The Latest Controversy over San Francisco Tech Company Buses." *SFGate,* July 23, 2021. https://www.sfgate.com/tech/article/san-francisco-tech-shuttle-buses-google-muni-16335538.php.

Gill, Rosalind. 2008. "Culture and Subjectivity in Neoliberal and Postfeminist Times." *Subjectivity* 25 (1): 432–445. https://doi.org/10.1057/sub.2008.28.

Gilmore, Ruth Wilson. 2007. *Golden Gulag: Prisons, Surplus, Crisis, and Opposition in Globalizing California.* Berkeley: University of California Press.

Giordano, Sara. 2014. "Scientific Reforms, Feminist Interventions, and the Politics of Knowing: An Auto-Ethnography of a Feminist Neuroscientist." *Hypatia* 29 (4): 755–773. https://doi.org/10.1111/hypa.12112.

———. 2016. "Building New Bioethical Practices through Feminist Pedagogies." *International Journal of Feminist Approaches to Bioethics* 9 (1): 81–103. https://doi.org/10.3138/ijfab.9.1.81.

———. 2017a. "Feminists Increasing Public Understandings of Science: A Feminist Approach to Developing Critical Science Literacy Skills." *Frontiers: A Journal of Women Studies* 38 (1): 100–123. https://doi.org/10.1353/fro.2017.a653262.

———. 2017b. "Those Who Can't, Teach: Critical Science Literacy as a Queer Science of Failure." *Catalyst: Feminism, Theory, Technoscience* 3 (1). https://doi.org/10.28968/cftt.v3i1.28790.

———. 2018. "Theorizing Feminist Tinkering with Science Methodologies." *Canadian Journal of Science, Mathematics and Technology Education* (18): 222–231. https://doi.org/10.1007/s42330-018-0027-y

———. 2020. "Feminist Science for the People: Feminist Approaches to Public Understanding of Science and Science Il/literacy." In *Handbook of Feminist Philosophy of Science*, edited by Sharon Crasnow and Kirsten Intemann, 250–260. New York: Routledge.

Giordano, Sara, and Yi-Lin Chung. 2018. "The Story Is That There Is No Story: Media Framing of Synthetic Biology and Its Ethical Implications in the *New York Times* (2005–2015)." *Journal of Science Communication* 17 (3): A02. https://doi.org/10.22323/2.17030202.

Giordano, Sara, and Maya Cruz. 2019. "Centering Reproductive Justice Praxis in the University Classroom through the Feminist Science Shop." Presentation at *Gender, Bodies, and Technology 2019: Technologics of Resistance*, Virginia Tech, Roanoke, VA, April 2019.

Goldman, Emma. 2011. *Living My Life (Two Volumes in One)*. New York: Cosimo Classics.

Gould, Stephen Jay. 1996. *The Mismeasure of Man*. New York: Norton.

Goven, Joanna. 2006. "Processes of Inclusion, Cultures of Calculation, Structures of Power Scientific Citizenship and the Royal Commission on Genetic Modification." *Science, Technology, and Human Values* 31 (5): 565–598. https://doi.org/10.1177/0162243906289612.

Grushkin, Daniel, Todd Kuiken, and Piers Millet. 2013. *Seven Myths and Realities about Do-It-Yourself Biology*. Washington, DC: Synthetic Biology Project and Woodrow Wilson International Center for Scholars. https://www.wilsoncenter.org/publication/seven-myths-and-realities-about-do-it-yourself-biology-0.

Gschmeidler, Brigitte, and Alexandra Seiringer. 2012. "'Knight in Shining Armour' or 'Frankenstein's Creation'? The Coverage of Synthetic Biology in German-Language Media." *Public Understanding of Science* 21 (2): 163–173. https://doi.org/10.1177/0963662511403876.

Halberstam, Judith (Jack). 2011. *The Queer Art of Failure*. Durham, NC: Duke University Press.

Hammonds, Evelynn, and Aimee Sands. 1993. "Never Meant to Survive: A Black Woman's Journey, An Interview with Evelynn Hammonds." In *The "Racial" Economy of Science: Toward a Democratic Future*, edited by Sandra Harding, 239–248. Bloomington: Indiana University Press.

Hammonds, Evelynn, and Banu Subramaniam. 2003. "A Conversation on Feminist Science Studies." *Signs: Journal of Women in Culture and Society* 28 (3): 923–944. https://doi.org/10.1086/345455.

Hamraie, Aimi. 2017. *Building Access: Universal Design and the Politics of Disability*. Minneapolis: University of Minnesota Press.

Haraway, Donna J. 1988. "Situated Knowledges: The Science Question in Feminism and the Privilege of Partial Perspective." *Feminist Studies* 14 (3): 575–599. https://doi.org/10.2307/3178066.

REFERENCES

——. 1994. "A Game of Cat's Cradle: Science Studies, Feminist Theory, Cultural Studies." *Configurations* 2 (1): 59–71. https://doi.org/10.1353/con.1994.0009.

——. 1997. *Modest_Witness@Second_Millennium.FemaleMan_Meets_OncoMouse: Feminism and Technoscience*. New York: Routledge.

——. 2008. *When Species Meet*. Minneapolis: University of Minnesota Press.

Harding, Sandra. 1987. "The Method Question." *Hypatia* 2 (3): 19–35. https://doi.org/10.1111/j.1527-2001.1987.tb01339.x.

——. 1991. *Whose Science? Whose Knowledge? Thinking from Women's Lives*. Ithaca, NY: Cornell University Press.

——, ed. 1993. *The "Racial" Economy of Science: Toward a Democratic Future*. Bloomington: Indiana University Press.

——. 2001. "After Absolute Neutrality: Expanding 'Science.'" In *Feminist Science Studies: A New Generation*, edited by Maralee Mayberry, Banu Subramaniam, and Lisa H. Weasel, 291–304. New York: Routledge.

——. 2004. *The Feminist Standpoint Theory Reader: Intellectual and Political Controversies*. New York: Routledge.

Harney, Stefano, and Fred Moten. 2013. *The Undercommons: Fugitive Planning and Black Study*. Wivenhoe, United Kingdom: Minor Compositions.

Hartsock, Nancy. 2004. "The Feminist Standpoint: Developing the Ground for a Specifically Feminist Historical Materialism." In *The Feminist Standpoint Theory Reader: Intellectual and Political Controversies*, edited by Sandra G. Harding, 35–54. New York: Routledge.

Helmreich, Stefan. 2008. "Species of Biocapital." *Science as Culture* 17 (4): 463–478. https://doi.org/10.1080/09505430802519256.

Hemmings, Clare. 2012. "Affective Solidarity: Feminist Reflexivity and Political Transformation." *Feminist Theory* 13 (2): 147–161. https://doi.org/10.1177/1464700112442643.

Herzig, Rebecca, and Banu Subramaniam. 2017. "Labor in the Age of 'Bio-Everything.'" *Radical History Review*, no. 127: 103–124. https://doi.org/10.1215/01636545-3690894.

Hess, Amanda. 2021. "'In This House' Yard Signs, and Their Curious Power." *New York Times*, October 29, 2021. https://www.nytimes.com/2021/10/29/arts/in-this-house-yard-signs.html.

Higgins, Marc, and Sara Tolbert. 2018. "A Syllabus for Response-Able Inheritance in Science Education." *Parallax* 24 (3): 273–294. https://doi.org/10.1080/13534645.2018.1496579.

Higgins, Shannon, Gord Wesmacott, and Julian Uzielli. 2015. "Alarm over Ahmed Mohamed's Clock Reveals Tinkering Tension." CBC Radio, September 18, 2015. https://www.cbc.ca/radio/thecurrent/the-current-for-september-18-2015-1.3233467/alarm-over-ahmed-mohamed-s-clock-reveals-tinkering-tension-1.3233474.

Himanen, Pekka. 2001. *The Hacker Ethic and the Spirit of the Information Age*. New York: Random House.

Hong, Grace Kyungwon. 2015a. *Death beyond Disavowal: The Impossible Politics of Difference*. Minneapolis: University of Minnesota Press.

——. 2015b. "Neoliberalism." *Critical Ethnic Studies* 1 (1): 56–67. https://doi.org/10.5749/jcritethnstud.1.1.0056.

hooks, bell. 1991. "Theory as Liberatory Practice." *Yale Journal of Law and Feminism* 4 (1): 1–12. http://hdl.handle.net/20.500.13051/7151.

Horne, Rebecca M., Matthew D. Johnson, Nancy L. Galambos, and Harvey J. Krahn. 2018. "Time, Money, or Gender? Predictors of the Division of Household Labour across Life Stages." *Sex Roles* 78 (11): 731–743. https://doi.org/10.1007/s11199-017-0832-1.

Hubbard, Ruth, Sandra Harding, Nancy Tuana, Sue V. Rosser, and Anne Fausto-Sterling. 1993. "Comments on Anne Fausto-Sterling's 'Building Two-Way Streets' [with Response]." *NWSA Journal* 5 (1): 45–81. http://www.jstor.org/stable/4316239.

Hubbard, Ruth, and Elijah Wald. 1999. *Exploding the Gene Myth: How Genetic Information Is Produced and Manipulated by Scientists, Physicians, Employers, Insurance Companies, Educators, and Law Enforcers.* Boston: Beacon Press. First published in 1993.

Ikemoto, Lisa C. 2017. "DIY Bio: Hacking Life in Biotech's Backyard." *UC Davis Law Review* 51 (2): 539–568. https://lawreview.law.ucdavis.edu/archives/51/2/diy-bio-hacking-life-biotechs-backyard.

Irani, Lilly. 2015. "Hackathons and the Making of Entrepreneurial Citizenship." *Science, Technology, and Human Values* 40 (5): 799–824. https://doi.org/10.1177/0162243915578486.

Irni, Sari. 2013. "The Politics of Materiality: Affective Encounters in a Transdisciplinary Debate." *European Journal of Women's Studies* 20 (4): 347–360. https://doi.org/10.1177/1350506812472669.

Irwin, Alan, and Brian Wynne. 1996. *Misunderstanding Science? The Public Reconstruction of Science and Technology.* Cambridge: Cambridge University Press.

Jacob, François. 1977. "Evolution and Tinkering." *Science* 196 (4295): 1161–1166. https://doi.org/10.1126/science.860134.

Jasanoff, Sheila. 2003. "Technologies of Humility: Citizen Participation in Governing Science." *Minerva* 41 (3): 223–244. https://doi.org/10.1023/A:1025557512320.

———. 2004. "The Idiom of Co-Production." In *States of Knowledge: The Co-Production of Science and Social Order,* edited by Sheila Jasanoff, 1–12. New York: Routledge.

Jen, Clare. 2015. "Do-It-Yourself Biology, Garage Biology, and Kitchen Science." *Knowing New Biotechnologies: Social Aspects of Technological Convergence,* edited by Matthias Wienroth and Eugenia Rodrigues, 125–141. New York: Routledge.

Jorgensen, Ellen D., and Daniel Grushkin. 2011. "Engage with, Don't Fear, Community Labs." *Nature Medicine* 17 (4): 411. https://doi.org/10.1038/nm0411-411.

Kaebnick, Gregory E. 2014. "Synthetic Biology." In *Handbook of Global Bioethics,* edited by Henk A.M.J. ten Have and Bert Gordijn, 811–26. Dordrecht, the Netherlands: Springer.

Katznelson, Ira. 2006. "New Deal, Raw Deal." *Souls* 8 (1): 9–11. https://doi.org/10.1080/10999940500516959.

Kazi, Nazia. 2015. "Ahmed Mohamed and the Imperial Necessity of Islamophilia." *Islamophobia Studies Journal* 3 (1): 115–126. https://doi.org/10.13169/islastudj.3.1.0115.

Kean, Sam. 2011. "A Lab of Their Own." *Science* 333 (6047): 1240–1241. https://doi.org/10.1126/science.333.6047.1240.

Keller, Evelyn Fox. 1977. "The Anomaly of a Woman in Physics." In *Working It Out: 23 Women, Writers, Scientists and Scholars Talk about Their Lives,* edited by Sara Ruddick and Pamela Daniels, 77–91. New York: Pantheon.

———. 1983. *A Feeling for the Organism: The Life and Work of Barbara McClintock.* San Francisco: W. H. Freeman.

———. 1995. *Reflections on Gender and Science.* New Haven, CT: Yale University Press. First published in 1985.

Kelly, Jean P. 2009. "Not So Revolutionary after All: The Role of Reinforcing Frames in US Magazine Discourse about Microcomputers." *New Media and Society* 11 (1–2): 31–52. https://doi.org/10.1177/1461444808100159.

Kelly, Susan E. 2003. "Public Bioethics and Publics: Consensus, Boundaries, and Participation in Biomedical Science Policy." *Science, Technology, and Human Values* 28 (3): 339–364. https://doi.org/10.1177/0162243903028003001.

Kelty, Christopher M. 2008. *Two Bits: The Cultural Significance of Free Software.* Durham, NC: Duke University Press. https://doi.org /10.1215/9780822389002.

REFERENCES

―――. 2010. "Outlaw, Hackers, Victorian Amateurs: Diagnosing Public Participation in the Life Sciences Today." *Journal of Science Communication* 9 (1) March: C03. https://doi.org/10.22323/2.09010303.

Keulartz, Jozef, and Henk van den Belt. 2016. "DIY-Bio—Economic, Epistemological and Ethical Implications and Ambivalences." *Life Sciences, Society and Policy* 12 (1): 7. https://doi.org/10.1186/s40504-016-0039-1.

King, Samantha. 2006. *Pink Ribbons, Inc.: Breast Cancer and the Politics of Philanthropy.* Minneapolis: University of Minnesota Press.

Kirchgasler, Kathryn L. 2019. "Strange Precipitate: How Interest in Science Produces Different Kinds of Students." In *STEM of Desire: Queer Theories and Science Education*, edited by Will Letts and Steve Fifield, 191–208. Leiden, the Netherlands: Brill.

Kirkpatrick, Graeme. 2002. "The Hacker Ethic and the Spirit of the Information Age." *Max Weber Studies* 2 (2): 163–185. http://www.jstor.org/stable/24579606.

Kisner, Jordan. 2021. "The Lockdown Showed How the Economy Exploits Women. She Already Knew." *New York Times*, February 17, 2021. https://www.nytimes.com/2021/02/17/magazine/waged-housework.html.

Korff, Gottfried. 1992. "From Brotherly Handshake to Militant Clenched Fist: On Political Metaphors for the Worker's Hand." *International Labor and Working-Class History* 42 (Fall): 70–81. https://doi.org/10.1017/S0147547900011236.

Kourany, Janet A. 2010. *Philosophy of Science after Feminism.* New York: Oxford University Press.

Kuhn, Thomas S. 1962. *The Structure of Scientific Revolutions.* Chicago: University of Chicago Press.

Kuznetsov, Stacey, Alex S. Taylor, Tim Regan, Nicolas Villar, and Eric Paulos. 2012. "At the Seams: DIYbio and Opportunities for HCI." In *DIS '12: Proceedings of the Designing Interactive Systems Conference*, 258–267. New York: Association for Computing Machinery.

Lane, Emily. 2016. "Alton Sterling and His CD-Selling Gig Made Him a Neighborhood Fixture." *NOLA.com*, July 7, 2016. https://www.nola.com/news/crime_police/article_314d96b0-3b3c-5df5-a924-56a9d9769404.html.

la paperson. 2017. *A Third University Is Possible: Uncovering the Decolonizing Ghost in the Colonizing Machine.* Minneapolis: University of Minnesota Press.

Lather, Patti. 2007. *Getting Lost: Feminist Efforts toward a Double(d) Science.* Albany: State University of New York Press.

Latour, Bruno. 1983. "Give Me a Laboratory and I Will Raise the World." In *Science Observed: Perspectives on the Social Study of Science*, edited by Karin D. Knorr-Cetina and Michael Mulkay, 141–170. London: Sage.

Le Doeuff, Michele. 1991. *Hipparchia's Choice: An Essay Concerning Women, Philosophy, Etc.* Translated by Trista Selous. Oxford: Blackwell.

Levine, Ross, and Yona Rubinstein. 2013. "Smart and Illicit: Who Becomes an Entrepreneur and Do They Earn More?" Working Paper 19276, National Bureau of Economic Research, Cambridge, MA, August 2013. https://doi.org/10.3386/w19276.

Lévi-Strauss, Claude. 1966. *The Savage Mind.* Chicago: University of Chicago Press.

Lewenstein, Bruce V. 1992. "The Meaning of 'Public Understanding of Science' in the United States after World War II." *Public Understanding of Science* 1 (1): 45–68. https://doi.org/10.1088/0963-6625/1/1/009.

―――. 2003. "Models of Public Communication of Science and Technology." Working paper, Cornell University, Ithaca, NY, June 2003. https://ecommons.cornell.edu/items/601f5747-d07a-4a52-a61d-d2fa8a7235bd.

Lipinski, Jed. 2011. "On Flatbush Avenue, Seven Stories Full of Ideas." *New York Times*, January 11, 2011. https://www.nytimes.com/2011/01/12/realestate/commercial/12incubate.html.

Long, Dayna. 2020. "'Enough Was Enough': The Science behind the Movement." *Red Madison* (blog), June 25, 2020. https://redmadison.com/2020/06/25/enough-was-enough-the-science-behind-the-movement/.

Longino, Helen E. 1990. *Science as Social Knowledge: Values and Objectivity in Scientific Inquiry*. Princeton, NJ: Princeton University Press.

Longino, Helen E., and Evelynn Hammonds. 1990. "Conflicts and Tensions in the Feminist Study of Gender and Science." In *Conflicts in Feminism*, edited by Marianne Hirsch and Evelyn Fox Keller, 164–183. New York: Routledge.

Lorde, Audre. 1984. "Learning from the 60s." In *Sister Outsider: Essays & Speeches*, 134–144. Berkeley: Crossing Press.

———. 1997. *The Cancer Journals*. Special edition. San Francisco: Aunt Lute.

Lövbrand, Eva, Roger Pielke, and Silke Beck. 2011. "A Democracy Paradox in Studies of Science and Technology." *Science, Technology, and Human Values* 36 (4): 474–496. https://doi.org/10.1177/0162243910366154.

Lowe, Lisa. 2015. *The Intimacies of Four Continents*. Durham, NC: Duke University Press.

Lugones, María. 1987. "Playfulness, 'World'-Travelling, and Loving Perception." *Hypatia* 2 (2): 3–19. http://www.jstor.org/stable/3810013.

———. 2003. *Pilgrimages/Peregrinajes: Theorizing Coalition against Multiple Oppressions*. Lanham, MD: Rowman & Littlefield.

———. 2007. "Heterosexualism and the Colonial/Modern Gender System." *Hypatia* 22 (1): 186–219. https://doi.org/10.1111/j.1527-2001.2007.tb01156.x.

Maher, Frances A., and Mary Kay T. Tetreault. 1994. *The Feminist Classroom*. New York: Basic Books.

Malatino, Hilary. 2017. "Biohacking Gender: Cyborgs, Coloniality, and the Pharmacopornographic Era." *Angelaki* 22 (2): 179–190. https://doi.org/10.1080/0969725X.2017.1322836.

Maldonado-Torres, Nelson. 2016. "10 Theses on Coloniality and Decoloniality." Lectures at Decolonizing Knowledge and Power: Postcolonial Studies, Decolonial Horizons Summer School, Barcelona, Catalonia, Spain, July 11–21.

Mamo, Laura, and Jennifer R. Fishman. 2013. "Why Justice? Introduction to the Special Issue on Entanglements of Science, Ethics, and Justice." *Science, Technology, and Human Values* 38 (2): 159–175. https://doi.org/10.1177/0162243912473162.

Martin, Emily. 2001. *The Woman in the Body: A Cultural Analysis of Reproduction*. Boston: Beacon.

Masters, Adam Stark. 2018. "How Making and Maker Spaces Have Contributed to Diversity and Inclusion in Engineering: A [Non-Traditional] Literature Review." Paper presented at 2018 CoNECD: The Collaborative Network for Engineering and Computing Diversity Conference, Crystal City, Virginia, April 2018. https://peer.asee.org/how-making-and-maker-spaces-have-contributed-to-diversity-and-inclusion-in-engineering-a-non-traditional-literature-review.

Maxigas. 2012. "Hacklabs and Hackerspaces: Tracing Two Genealogies." *Journal of Peer Production*, no. 2. http://peerproduction.net/issues/issue-2/peer-reviewed-papers/hacklabs-and-hackerspaces/.

Mayberry, Maralee, and Margaret N. Rees. 1999. "Feminist Pedagogy, Interdisciplinary Praxis and Science Education." In *Meeting the Challenge: Innovative Feminist Pedagogies in Action*, edited by Maralee Mayberry and Ellen Cronan Rose, 193–214. New York: Routledge.

Mayberry, Maralee, Banu Subramaniam, and Lisa H. Weasel. 2001. *Feminist Science Studies: A New Generation*. New York: Routledge.

McDonald, Terrence J. 1996. *The Historic Turn in the Human Sciences*. Ann Arbor: University of Michigan Press.

REFERENCES

McIntosh, Peggy. 2004. "White Privilege: Unpacking the Invisible Knapsack." In *Race, Class, and Gender in the United States: An Integrated Study*, 6th ed., edited by Paula S. Rothenberg, 188–192. New York: Worth.

McKittrick, Katherine. 2021. *Dear Science and Other Stories*. Durham, NC: Duke University Press.

Melamed, Jodi. 2006. "The Spirit of Neoliberalism: From Racial Liberalism to Neoliberal Multiculturalism." *Social Text* 24 (4): 1–24. https://doi.org/10.1215/01642472-2006-009.

———. 2015. "Racial Capitalism." *Critical Ethnic Studies* 1 (1): 76–85. https://doi.org/10.5749/jcritethnstud.1.1.0076.

Meyer, Morgan. 2013a. "Assembling, Governing, and Debating an Emerging Science: The Rise of Synthetic Biology in France." *BioScience* 63 (5): 373–379. https://doi.org/10.1525/bio.2013.63.5.10.

———. 2013b. "Domesticating and Democratizing Science: A Geography of Do-It-Yourself Biology." *Journal of Material Culture* 18 (2): 117–134. https://doi.org/10.1177/1359183513483912.

———. 2016. "Steve Jobs, Terrorists, Gentlemen, and Punks: Tracing Strange Comparisons of Biohackers." In *Practising Comparison: Logics, Relations, Collaborations*, edited by Joe Deville, Michael Guggenheim, and Zuzana Hrdličková, 281–305. Manchester, United Kingdom: Mattering.

Meyer, Morgan, and Frédéric Vergnaud. 2020. "The Rise of Biohacking: Tracing the Emergence and Evolution of DIY Biology through Online Discussions." *Technological Forecasting and Social Change* 160 (November): 120206. https://doi.org/10.1016/j.techfore.2020.120206.

Meyerhoff, Eli. 2019. *Beyond Education: Radical Studying for Another World*. Minneapolis: University of Minnesota Press.

Michael, Chris, and Ellie Violet Bramley. 2014. "Spike Lee's Gentrification Rant— Transcript: 'Fort Greene Park Is Like the Westminster Dog Show.'" *The Guardian*, February 26, 2014. https://www.theguardian.com/cities/2014/feb/26/spike-lee-gentrification-rant-transcript.

Mignolo, Walter D. 2015. "Sylvia Wynter: What Does It Mean to Be Human?" In *Sylvia Wynter: On Being Human as Praxis*, edited by Katherine McKittrick, 106–123. Durham, NC: Duke University Press.

Minkler, Meredith, and Nina Wallerstein, eds. 2011. *Community-Based Participatory Research for Health: From Process to Outcomes*. San Francisco: Jossey-Bass.

Mody, Cyrus C. M. 2006. "Corporations, Universities, and Instrumental Communities: Commercializing Probe Microscopy, 1981–1996." *Technology and Culture* 47 (1): 56–80. https://doi.org/10.1353/tech.2006.0085.

Moore, Jason W. 2016. *Anthropocene or Capitalocene? Nature, History, and the Crisis of Capitalism*. Oakland, CA: PM Press.

Mullis, Kary B. 1993. "Nobel Lecture, December 8, 1993: The Polymerase Chain Reaction." In *Nobel Lectures, Chemistry 1991–1995*, edited by Bo G. Malmström. Singapore: World Scientific, 1997. Republished on The Nobel Prize, https://www.nobelprize.org/prizes/chemistry/1993/mullis/lecture/.

Murch, Donna Jean. 2010. *Living for the City: Migration, Education, and the Rise of the Black Panther Party in Oakland, California*. Chapel Hill: University of North Carolina Press.

Murphy, Michelle. 2012. *Seizing the Means of Reproduction: Entanglements of Feminism, Health, and Technoscience*. Durham, NC: Duke University Press.

Musil, Caryn McTighe, ed. 2001. *Gender, Science, and the Undergraduate Curriculum: Building Two-Way Streets*. Washington, DC: Association of American Colleges and Universities.

Nair, Prashant. 2012. "Profile of George M. Church." *Proceedings of the National Academy of Sciences of the United States of America* 109 (30): 11893–11895. https://doi.org/10.1073/pnas.1204148109.

Narek, Diane. 1970. "A Woman Scientist Speaks." In *Voices from Women's Liberation*, edited by Leslie B. Tanner, 325–329. New York: Mentor.

National Academies of Sciences, Engineering, and Medicine. 2017. *Preparing for Future Products of Biotechnology*. Washington, DC: National Academies Press. https://doi.org/10.17226/24605.

National Center for Science and Engineering Statistics. 2021. *Women, Minorities, and Persons with Disabilities in Science and Engineering*. Alexandria, VA: National Science Foundation. https://ncses.nsf.gov/pubs/nsf21321/.

Nelson, Alondra. 2011. *Body and Soul: The Black Panther Party and the Fight against Medical Discrimination*. Minneapolis: University of Minnesota Press.

———. 2016. *The Social Life of DNA: Race, Reparations, and Reconciliation after the Genome*. Boston: Beacon.

Newfield, Christopher. 2008. *Unmaking the Public University: The Forty-Year Assault on the Middle Class*. Cambridge, MA: Harvard University Press.

Novella, Caro. 2017. "Arousing Formlessness: Collaborative Encounters with Chemotherapy." *Performance Research* 22 (6): 54–56. https://doi.org/10.1080/13528165.2017.1414405.

———. 2021. "On_co-Creations: Rehearsing Cancer Ecologies, Making Cancer Multiple in Art/Activism." PhD diss., University of California, Davis.

———. 2022. "What Are We Waiting For? Rehearsal as Arousal Politics in Oncogrrrls." *The Drama Review* 66 (2) June: 101–114. https://www.muse.jhu.edu/article/858118.

Ong, Aihwa. 2006. *Neoliberalism as Exception: Mutations in Citizenship and Sovereignty*. Durham, NC: Duke University Press.

Patel, Leigh. 2015. *Decolonizing Educational Research: From Ownership to Answerability*. New York: Routledge.

Patterson, Meredith L. 2010. "A Biopunk Manifesto." LiveJournal, January 30, 2010. https://maradydd.livejournal.com/496085.html.

Pollock, Anne, and Banu Subramaniam. 2016. "Resisting Power, Retooling Justice Promises of Feminist Postcolonial Technosciences." *Science, Technology, and Human Values* 41 (6): 951–966. https://doi.org/10.1177/0162243916657879.

Povinelli, Elizabeth A. 2016. *Geontologies: A Requiem to Late Liberalism*. Durham, NC: Duke University Press.

Presidential Commission for the Study of Bioethical Issues. 2010. *New Directions: The Ethics of Synthetic Biology and Emerging Technologies*. December 2010. https://bioethicsarchive.georgetown.edu/pcsbi/sites/default/files/PCSBI-Synthetic-Biology-Report-12.16.10_0.pdf.

Quijano, Aníbal. 2000. "Coloniality of Power and Eurocentrism in Latin America." *International Sociology* 15 (2): 215–232. https://doi.org/10.1177/0268580900015002005.

Rabinow, Paul, and Gaymon Bennett. 2009. "Synthetic Biology: Ethical Ramifications 2009." *Systems and Synthetic Biology* 3 (1–4): 99–108. https://doi.org/10.1007/s11693-009-9042-7.

Rajan, Kaushik Sunder. 2006. *Biocapital: The Constitution of Postgenomic Life*. Durham, NC: Duke University Press.

Reardon, Jenny. 2013. "On the Emergence of Science and Justice." *Science, Technology, and Human Values* 38 (2): 176–200. https://doi.org/10.1177/0162243912473161.

Reeves, Richard V. 2017. *Dream Hoarders: How the American Upper Middle Class Is Leaving Everyone Else in the Dust, Why That Is a Problem, and What to Do about It*. Washington, DC: Brookings Institution.

REFERENCES

Riley, Donna. 2017. "Rigor/Us: Building Boundaries and Disciplining Diversity with Standards of Merit." *Engineering Studies* 9 (3): 249–265. https://doi.org/10.1080/19378629.2017.1408631.

Roberts, Dorothy. 1997. *Killing the Black Body: Race, Reproduction, and the Meaning of Liberty.* New York: Vintage.

Robinson, Cedric J. 1983. *Black Marxism: The Making of the Black Radical Tradition.* London: Zed.

Rojas, Lucia Egaña. 2013. "Notes Towards a Transfeminist Technology." *Root*, December 18, 2013. https://word.root.ps/?p=469.

Roosth, Sophia. 2013. "Biobricks and Crocheted Coral: Dispatches from the Life Sciences in the Age of Fabrication." *Science in Context* 26 (1): 153–171. https://doi.org/10.1017/S0269889712000324.

———. 2017. *Synthetic: How Life Got Made.* Chicago: University of Chicago Press.

Rosalsky, Greg. 2021. "How California Homelessness Became a Crisis." *NPR*, June 8, 2021. https://www.npr.org/sections/money/2021/06/08/1003982733/squalor-behind-the-golden-gate-confronting-californias-homelessness-crisis.

Rose, Hilary. 1992. "Gendered Reflexions on the Laboratory in Medicine." In *The Laboratory Revolution in Medicine*, edited by Andrew Cunningham and Perry Williams, 324–342. Cambridge: Cambridge University Press.

Roth, Wolff-Michael, and Angela Calabrese Barton. 2004. *Rethinking Scientific Literacy.* New York: Routledge Falmer.

Roy, Deboleena. 2004. "Feminist Theory in Science: Working toward a Practical Transformation." *Hypatia* 19 (1): 255–279. https://doi.org/10.1111/j.1527-2001.2004.tb01277.x.

———. 2008. "Asking Different Questions: Feminist Practices for the Natural Sciences." *Hypatia* 23 (4):134–157. doi:10.1111/j.1527-2001.2008.tb01437.x

———. 2012. "Feminist Approaches to Inquiry in the Natural Sciences: Practices for the Lab." In *Handbook of Feminist Research: Theory and Praxis*, edited by Sharlene Nagy Hesse-Biber, 313–330. Thousand Oaks, CA: Sage.

Rusert, Britt. 2017. *Fugitive Science: Empiricism and Freedom in Early African American Culture.* New York: New York University Press.

Sandoval, Chela. 2000. *Methodology of the Oppressed.* Minneapolis: University of Minnesota Press.

Schiebinger, Londa L. 2004. *Nature's Body: Gender in the Making of Modern Science.* New Brunswick, NJ: Rutgers University Press.

Shapin, Steven. 1991. "'A Scholar and a Gentleman': The Problematic Identity of the Scientific Practitioner in Early Modern England." *History of Science* 29 (3): 279–327. https://doi.org/10.1177/007327539102900303.

Singh, Vineeta. 2021. "'Never Waste a Good Crisis': Critical University Studies during and after a Pandemic." *American Quarterly* 73 (1): 181–193. https://doi.org/10.1353/aq.2021.0014.

Slaton, Amy E. 2021. "Racial Capitalism and the Making of Talent: STEM Achievement and the Making of Human Difference in the Twenty-First Century." Presented at the Virginia Tech STS Seminar, Virginia Tech, October 14.

Slaton, Amy E., Erin A. Cech, and Donna M. Riley. 2019. "Yearning, Learning, and Earning: The Gritty Ontologies of American Engineering Education." In *STEM of Desire: Queer Theories and Science Education*, edited by Will Letts and Steve Fifield, 319–340. Leiden, the Netherlands: Brill.

Smith, David R. 2014. "The Outsourcing and Commercialization of Science." *EMBO Reports* 16 (1): 14–16. https://doi.org/10.15252/embr.201439672.

Spade, Dean. 2020. *Mutual Aid: Building Solidarity during This Crisis (and the Next).* London: Verso.

Spanier, Bonnie B. 1986. "Women's Studies and the Natural Sciences: A Decade of Change." *Frontiers: A Journal of Women Studies* 8 (3): 66–72. https://doi.org/10.2307/3346375.

Spencer, Robyn C. 2016. *The Revolution Has Come: Black Power, Gender, and the Black Panther Party in Oakland.* Durham, NC: Duke University Press.

Spivak, Gayatri Chakravorty. 1976. "Translator's Preface." In *Of Grammatology,* Jacques Derrida, translated by Gayatri Chakravorty Spivak, ix–lxxxviii. Baltimore, MD: Johns Hopkins University Press.

———. 1994. "Responsibility." *Boundary 2* 21, no. 3 (Autumn): 19–64. https://doi.org/10.2307/303600.

Stacey, Judith. 2004. "Cruising to Familyland: Gay Hypergamy and Rainbow Kinship." *Current Sociology* 52 (2): 181–197. https://doi.org/10.1177/0011392104041807.

Stamboliyska, Rayna. 2012. "DIY Science—How Do We Make DIYBio Sustainable?" *SpotOn: Science Policy, Outreach and Tools Online* (blog), December 13, 2012. https://web.archive.org/web/20130223040004/http://www.nature.com/spoton/2012/12/spoton-nyc-diy-science-how-do-we-make-diybio-sustainable/.

Stein, Sharon, Vanessa Andreotti, Rene Suša, Sarah Amsler, Dallas Hunt, Cash Ahenakew, Elwood Jimmy, Tereza Cajkova, Will Valley, and Camilla Cardoso. 2020. "Gesturing towards Decolonial Futures: Reflections on Our Learnings Thus Far." *Nordic Journal of Comparative and International Education* 4 (1): 43–65. https://doi.org/10.7577/njcie.3518.

Stilgoe, Jack, Simon J. Lock, and James Wilsdon. 2014. "Why Should We Promote Public Engagement with Science?" *Public Understanding of Science* 23 (1): 4–15. https://doi.org/10.1177/0963662513518154.

Subramaniam, Banu. 2005. "Laboratories of Our Own: New Productions of Gender and Science." In *Women's Studies for the Future: Foundations, Interrogations, Politics,* edited by Elizabeth Lapovsky Kennedy and Agatha Beins, 229–242. New Brunswick, NJ: Rutgers University Press.

———. 2009. "Moored Metamorphoses: A Retrospective Essay on Feminist Science Studies." *Signs: Journal of Women in Culture and Society* 34 (4): 951–980. https://doi.org/10.1086/597147.

Talbot, Margaret. 2020. "The Rogue Experimenters." *New Yorker,* May 18, 2020. https://www.newyorker.com/magazine/2020/05/25/the-rogue-experimenters.

TallBear, Kim. 2013. *Native American DNA: Tribal Belonging and the False Promise of Genetic Science.* Minneapolis: University of Minnesota Press.

Temple, James. 2014. "The Time Traveler: George Church Plans to Bring Back a Creature That Went Extinct 4,000 Years Ago." *Vox,* December 8, 2014. https://www.vox.com/2014/12/8/11633602/the-time-traveler-george-church-is-racing-into-the-future-and.

Tiede, Hans-Joerg, Samantha McCarthy, Isaac Kamola, and Alyson K. Spurgas. 2021. "Data Snapshot: Whom Does Campus Reform Target and What Are the Effects?" *Academe* 107 (2). https://www.aaup.org/article/data-snapshot-whom-does-campus-reform-target-and-what-are-effects.

Tocchetti, Sara. 2012. "DIYbiologists as 'Makers' of Personal Biologies: How MAKE Magazine and Maker Faires Contribute in Constituting Biology as a Personal Technology." *Journal of Peer Production,* no. 2: 1–9. http://peerproduction.net/issues/issue-2/peer-reviewed-papers/diybiologists-as-makers/.

———. 2014. "How Did DNA Become Hackable and Biology Personal? Tracing the Self-Fashioning of the DIYBio Network." PhD diss., London School of Economics and Political Science.

Tocchetti, Sara, and Sara Angeli Aguiton. 2015. "Is an FBI Agent a DIY Biologist Like Any Other? A Cultural Analysis of a Biosecurity Risk." *Science, Technology, and Human Values* 40 (5): 825–853. https://doi.org/10.1177/0162243915589634.

REFERENCES

Toupin, Sophie. 2014. "Feminist Hackerspaces: The Synthesis of Feminist and Hacker Cultures." *Journal of Peer Production*, no. 5. http://peerproduction.net/editsuite/issues/issue-5-shared-machine-shops/peer-reviewed-articles/feminist-hackerspaces-the-synthesis-of-feminist-and-hacker-cultures/.

Traweek, Sharon. 2009. *Beamtimes and Lifetimes: The World of High Energy Physicists.* Cambridge, MA: Harvard University Press.

Turkle, Sherry. 2005. *The Second Self: Computers and the Human Spirit.* Cambridge, MA: MIT Press.

Tutton, Richard. 2007. "Constructing Participation in Genetic Databases: Citizenship, Governance, and Ambivalence." *Science, Technology, and Human Values* 32 (2): 172–195. https://doi.org/10.1177/0162243906296853.

Tzanelli, Rodanthi, and Majid Yar. 2016. "Breaking Bad, Making Good: Notes on a Televisual Tourist Industry." *Mobilities* 11 (2): 188–206. https://doi.org/10.1080/17450101.2014.929256.

Ureña, Carolyn. 2017. "Loving from Below: Of (De)Colonial Love and Other Demons." *Hypatia* 32 (1): 86–102. https://doi.org/10.1111/hypa.12302.

Van der Haak, Bregtje, dir. 2012. *DNA Dreams* [documentary]. Hilversum, the Netherlands: Vrijzinnig Protestantse Radio Omroep (VPRO).

Vinsel, Lee, and Andrew L. Russell. 2020. *The Innovation Delusion: How Our Obsession with the New Has Disrupted the Work That Matters Most.* New York: Currency.

Visperas, Cristina, Kimberly Juanita Brown, and Jared Sexton. 2016. "Introduction." *Catalyst: Feminism, Theory, Technoscience* 2 (2): 1–12. https://doi.org/10.28968/cftt.v2i2.28797.

Wachelder, Joseph. 2003. "Democratizing Science: Various Routes and Visions of Dutch Science Shops." *Science, Technology, and Human Values* 28 (2): 244–273. https://doi.org/10.1177/0162243902250906.

Wallerstein, Immanuel Maurice. 2004. *The Uncertainties of Knowledge.* Philadelphia: Temple University Press.

Warner, Michael. 2002. "Publics and Counterpublics." *Public Culture* 14 (1): 49–90. https://muse.jhu.edu/article/26277.

Washington, Harriet A. 2006. *Medical Apartheid: The Dark History of Medical Experimentation on Black Americans from Colonial Times to the Present.* New York: Doubleday.

Weasel, Lisa H. 2001. "Laboratories without Walls: The Science Shop as a Model for Feminist Community Science in Action." In *Feminist Science Studies: A New Generation*, edited by Maralee Mayberry, Banu Subramaniam, and Lisa H. Weasel, 305–320. New York: Routledge.

Weiner, Charles. 2001. "Drawing the Line in Genetic Engineering: Self-Regulation and Public Participation." *Perspectives in Biology and Medicine* 44 (2): 208–220. https://doi.org/10.1353/pbm.2001.0039.

Weinstein, Matthew. 2010. "A Science Literacy of Love and Rage: Identifying Science Inscription in Lives of Resistance." *Canadian Journal of Science, Mathematics and Technology Education* 10 (3): 267–277. https://doi.org/10.1080/14926156.2010.504489.

Weisheit, Ralph A., and L. Edward Wells. 2010. "Methamphetamine Laboratories: The Geography of Drug Production." *Western Criminology Review* 11 (2): 9–26. http://wcr.sonoma.edu/v11n2/Weisheit.pdf.

Whatley, Mariamne. 1986. "Taking Feminist Science to the Classroom: Where Do We Go from Here?" In *Feminist Approaches to Science*, edited by Ruth Bleier, 181–190. New York: Pergamon.

White, Paul. 2009. "Darwin's Emotions: The Scientific Self and the Sentiment of Objectivity." *Isis* 100 (4): 811–826. https://doi.org/10.1086/652021.

Wilkinson, Alissa. 2016. "Satan, the Pope, and Dungeons & Dragons: How Jack Chick's Cartoons Informed American Fundamentalism." *Vox*, November 8, 2016. https://www.vox.com/culture/2016/11/8/13426962/jack-chick-alt-right-fundamentalism-tracts-catholics-trump.

Willey, Angela. 2016a. "Biopossibility: A Queer Feminist Materialist Science Studies Manifesto, with Special Reference to the Question of Monogamous Behavior." *Signs: Journal of Women in Culture and Society* 41 (3): 553–577. https://doi.org/10.1086/684238.

———. 2016b. *Undoing Monogamy: The Politics of Science and the Possibilities of Biology.* Durham, NC: Duke University Press.

Willey, Angela, and Banu Subramaniam. 2017. "Inside the Social World of Asocials: White Nerd Masculinity, Science, and the Politics of Reverent Disdain." *Feminist Studies* 43 (1): 13–41. https://doi.org/10.15767/feministstudies.43.1.0013.

Williams, Lauren. 2014. "The Terrifying Racial Stereotypes Laced through Darren Wilson's Testimony." *Vox*, November 25, 2014. https://www.vox.com/2014/11/25/7283327/michael-brown-racist-stereotypes.

Wilson, Duncan. 2012. "Who Guards the Guardians? Ian Kennedy, Bioethics and the 'Ideology of Accountability' in British Medicine." *Social History of Medicine* 25 (1): 193–211. https://doi.org/10.1093/shm/hkr090.

Wisnioski, Matthew H., Eric S. Hintz, and Marie Stettler Kleine. 2019. *Does America Need More Innovators?* Cambridge, MA: MIT Press.

Wohlsen, Marcus. 2012. *Biopunk: Solving Biotech's Biggest Problems in Kitchens and Garages.* New York: Current.

Woolf, Virginia. 1929. *A Room of One's Own.* New York: Harcourt Brace & Co.

Wu, Diana Pei. 2013. "Eco Means Home: Race, Gender, Class and a New/Old Ecology." Lecture at San Diego State University, San Diego, CA, February 27.

Wynne, Brian. 2007. "Public Participation in Science and Technology: Performing and Obscuring a Political–Conceptual Category Mistake." *East Asian Science, Technology and Society* 1 (1): 99–110. https://doi.org/10.1215/s12280-007-9004-7.

Wynter, Sylvia. 1984. "The Ceremony Must Be Found: After Humanism." *Boundary 2*, 12/13 (3–1): 19–70. https://doi.org/10.2307/302808.

———. 1994. "No Humans Involved: An Open Letter to My Colleagues." *Forum N.H.I.: Knowledge for the 21st Century* 1 (1): 42–73. http://carmenkynard.org/wp-content/uploads/2013/07/No-Humans-Involved-An-Open-Letter-to-My-Colleagues-by-SYLVIA-WYNTER.pdf.

———. 2003. "Unsettling the Coloniality of Being/Power/Truth/Freedom: Towards the Human, after Man, Its Overrepresentation—an Argument." *CR: The New Centennial Review* 3 (3): 257–337. https://doi.org/10.1353/ncr.2004.0015.

Xu, Hongwei, and Martin Ruef. 2004. "The Myth of the Risk-Tolerant Entrepreneur." *Strategic Organization* 2 (4): 331–355. https://doi.org/10.1177/1476127004047617.

Zhang, Sarah. 2017. "Whatever Happened to the Glowing Plant Kickstarter?" *The Atlantic*, April 20, 2017. https://www.theatlantic.com/science/archive/2017/04/whatever-happened-to-the-glowing-plant-kickstarter/523551/.

Zimmerman, Tegan. 2012. "The Politics of Writing, Writing Politics: Virginia Woolf's A [Virtual] Room of One's Own." *Journal of Feminist Scholarship* 3:35–55. https://digitalcommons.uri.edu/jfs/vol3/iss3/4.

Index

ableism, 99, 184n2

academics, 18, 146, 173

accountability: of academics, 18; and democratic sciences, 82; DIY biology and, 89; and feminist ethics, 151; and feminist science, 134, 154–155, 166; and knowledge production, 18, 151, 160–161, 168–169; science and, 81, 82, 133; scientific ethics and, 86, 93; and structural power, 133; tinkerers and, 59, 69, 109. *See also* responsibility

activists, Black, 19, 164

activists, disability, 99, 143, 146, 149, 162–163, 166–167

activists, environmental, 5, 93

activists, feminist, 3

activists, health, 8, 139, 143, 147, 148, 164

activists, Indigenous, 94

activists, racial justice, 5

activists, reproductive justice, 142

activists, right-wing, 120

activists, transfeminist, 141, 152, 153

activists: as antiscience, 123; and assembly (concept), 144; and categories of difference, 146–147; and COVID-19, 171; and democratic sciences, 14, 139; and DIY biology communities, 31t; and DIY biology labs, 25; and economic power, 139; feminist scientists as, 142; and gendered racial capitalism, 8; and genetic engineering, 35; knowledge production by, 143, 162–163; and the open source movement, 67; and positionality, 156; and science, 1–2, 124; and standpoint theories, 146; and traditional science, 32. *See also* Black Panther Party

Adams, M, 6–7

affect: and biocitizenship, 15–16; and capitalism, 157; definition of, 107; and DIY biology, 165; and DIY biology communities, 17; and education, 60; and exclusion, 9; feminist scientists and, 109; as gendered, 41; and knowledge production, 164; and nature (natural world), 109; under neoliberalism, 9; science and, 15, 60, 108, 113, 114t, 156, 172; and science literacy, 14, 82, 147; and STEM (science, technology, engineering, and mathematics), 60; and tinkerers, 82; truth and, 113, 114t. *See also* emotions; passion for science

affective dissonance, 117, 163, 164

affective economies, 107

affective solidarity, 163

Afghanistan, 89

Africa, 79

Ahmed, Sara, 15, 107, 113

AIDS movements, 56, 139, 150

Allen, Caitlyn, 118

All Lives Matter, 93

Altman, Mitch, 180n13

amateurs, 11, 29, 30, 31t, 74, 90. *See also* tinkerers

America. *See* United States

American exceptionalism, 64

Amirav-Drory, Omri, 34

anarchism, 138

ancestry testing, 45, 47, 49, 50, 180n14

animal research, 153

anticapitalism, 11, 54, 121, 138, 140, 157, 171

anti-intellectualism, 11, 113

antiracism, 11, 49, 68, 123, 138, 139, 148–149

antiscience positions, 100, 113, 120, 123, 124

Anzaldúa, Gloria, 132–133, 163–164

Arabs, 79
Åsberg, Cecilia, 112
Asia, 79
Asilomar Conference on Recombinant
  DNA (1975), 85
assemblage (concept), 166, 172
assembly (concept), 140, 144, 166
Atanasoski, Neda, 66
Aguiton, Sara, 77
automation, 45, 46, 47, 66–67
Avatar (2009 film), 35

Baldwin, James, 158
Baltimore Underground Science Space
  (BUGSS), 25, 28–29, 31t, 57, 82, 83, 91
Barad, Karen, 82, 128, 131, 161, 168, 185n3
Barr, Jean, 115, 128, 142
Barres, Ben, 116
Bates College, 127
BDSM, 152
Benjamin, Ruha, 16, 88, 115, 123, 183n2
Bernstein, Robin, 63
beta galactosidase, 75
BGI (Chinese genomics institute), 50
Big Bio, 87t; automation by, 46; vs.
  democratic sciences, 94–95; vs. DIY
  biology, 67; gatekeeping by, 32; and
  scientific ethics, 81; and synthetic biology,
  84–85; tinkerers and, 13, 62; and truth,
  178n5. *See also* laboratories, traditional
BioArt Laboratories, 31t, 32, 84
BioBricks Foundation, 31t, 92
biocapital, 14
biocitizenship, 14, 15–16, 75, 78, 123, 183n2
BioCoder (magazine), 74
BioCurious, 29, 31t, 34, 58–59, 75, 84, 91
biodefection, 115, 123
biodefectors, 16, 183n2
bioeconomies, 14–15
bioerror, 74, 75, 79, 81, 81t, 82
biohackers, 62, 78, 90, 92, 137, 138, 140, 150
biohacking, 54, 138, 152, 155. *See also*
  biology, DIY; biology, synthetic
biological determinism, 8, 9, 59, 64, 69, 110,
  128, 147. *See also* genetics (of a person);
  racism
biological materials, 14, 67
Biologigaragen, 31t, 79, 84
biologists, 39–40, 46, 85
biologists, DIY: as ambassadors for science,
  80; claims of disenfranchisement by, 62;
  and concept of intelligence, 56–57;
  co-optation of social justice ideas by, 93;
  counterculture image of, 76, 77, 78, 90;
  and COVID-19, 171; and entrepreneurship, 62–63, 82, 159; and exclusion, 94;
  and the Federal Bureau of Investigation

(FBI), 27–28, 76–78; and gender, 65–66; as
  hackers, 76; and Homeland Security,
  76–77; and inclusion, 89; and informed
  consent, 73; invoking of revolution by,
  31–32; liminal status of, 78, 82, 89; and
  passion for science, 30; positionality of,
  76–77, 83; as the public, 86, 89, 92; public
  image of, 76–77; public relations work by,
  28–30, 79–80, 81, 82; qualifications to be,
  30; as revolutionaries, 80; roles of, 34; as
  safe, 82; and scientific ethics, 94; and
  social class, 76; vs. terrorists, 76, 78, 92; as
  tinkerers, 57, 90; and traditional science,
  82, 83, 90, 92. *See also* amateurs;
  biohackers; communities, DIY biology;
  scientists
biologists, synthetic, 130
biology, 7, 8, 14, 15, 56, 58
biology, DIY: vs. academic sciences, 89; and
  accountability, 89; and affect, 165; and
  amateurs, 90; and automation, 45, 46; vs.
  Big Bio, 67; bioluminescence project in,
  34–35; and biomedical field, 80; and
  capitalism, 48–49, 62–63; classes in, 39,
  45, 47, 79, 80, 165; and coloniality, 63; and
  concept of intelligence, 69; and control of
  knowledge, 46, 67; and DNA, 47; early
  history of, 27; and economic success, 17,
  20, 49, 66; and engineering, 57; and
  entrepreneurship, 17, 29, 34, 47, 48–49, 54,
  62–63, 68, 69; and exploration, 49, 82; and
  fear of science, 80, 81t, 82; and funding
  for science, 34, 42, 48, 61, 83; and
  gendered racial capitalism, 18, 50–51, 54,
  65, 68–69; and genetics (of a person), 48;
  and genomic research, 45, 84–85; and
  hacking, 54; and innovation, 17, 32, 49,
  58–59; and knowledge production, 34, 47;
  as labor, 45; and new democratic sciences,
  138; and passion for science, 58–59, 69; as
  play, 29, 44–45, 62–63, 82, 83; political and
  economic status of, 14–15; and politics of
  inclusion, 30, 94, 140; property rights
  and, 48–49; and the public, 75, 81t, 85;
  public access to, 29; and race, 65; and
  racialization, 92; as revolution, 62, 74, 78;
  rights discourse concerning, 84; and
  science literacy, 80; and scientific ethics,
  75–76, 85, 89, 94; and social class, 61–63;
  and synthetic biology, 84; as technoscience, 57; and terrorism, 75–76; vs.
  traditional science, 78, 80, 89; vs.
  universities and colleges, 67, 81; use of
  DNA in, 4, 48. *See also* biohacking
biology, synthetic: and Big Bio, 84–85;
  communities in, and feminist activists, 3;
  and computer tinkering, 61; and control

INDEX

of knowledge, 46; definition of, 2–3, 27, 45–46, 85; and DIY biology, 84; and gendered racial capitalism, 68–69; and genetic engineering, 27, 45–46, 85; and genetic research, 2–3, 46–47; history of, 84–85; lack of feminists in, 137; and nature (natural world), 46; news reports concerning, 85; as play, 45; and the public, 85; and scientific ethics, 35, 69, 74, 85; and social class, 61–63; and social scientists, 85–86; terrorism and, 78; use of DNA in, 46, 85. *See also* biohacking

bioluminescence, 34–35

biomedical community, 81, 82

biomedicine, 80, 81t

biopossibilities, 168

"Biopunk Manifesto, A" (Patterson), 83

Biospace, 31t, 32, 83

biotechnology, 92

Biotech without Borders, 92

bioterror, 74, 81, 81t, 82

biovalue, 14

Birch, Kean, 14

Birke, Lynda, 112, 115, 122, 128, 142

Black liberation movements, 146, 164

Black Lives Matter (BLM), 5, 18, 93

Black Panther Party, 8, 139, 146–147, 148, 149, 150, 166

Black people: and capitalism, 178n6; dehumanization of, 18; discrimination toward, 147; and entrepreneurship, 13; and eugenics, 55; as excluded from science, 100, 110–111, 123; families of, 9; health care and, 139; and politics of inclusion, 132; and self-determination, 148, 149, 164, 166; state violence toward, 19, 171

Black Power movement, 32, 62

Black studies, 8, 18, 122, 143

Bleier, Ruth, 103, 129

Bobe, Jason, 27

borders. *See* liminality

BOSLab (formerly BOSSLab), 29, 91–92

Bourcier, Sam, 141

Breaking Bad (television series), 77

breast cancer, 137, 139, 154

bricoleurs, 178n9

Brightwork, 31t

Brown, Wendy, 94

Brown people, 139

"Building Two-Way Streets: The Case for Feminism and Science" (Fausto-Sterling), 109, 111

Butler, Judith, 167

Calvert, Jane, 45, 85

Cancer Journals (Lorde), 139, 145

capital (class status), 8, 14

capitalism, gendered racial: and biological determinism, 69; and biology, 58; and categories of difference, 140; challenges to, 8, 133; and coloniality, 93; and COVID-19, 171; definition of, 6–7; and democratic sciences, 89, 91; and disenfranchisement, 8, 131; and distribution of resources, 162; and DIY biology, 18, 50–51, 54, 65, 68–69; and economic power, 58, 93; and exclusion, 58, 69, 140; and feminism, 140; and inclusion, 140; and knowledge production, 58, 65; and medicine, 8; and people of color, 93; and politics of inclusion, 140; the public and, 86, 91; and synthetic biology, 68–69; and systemic inequalities, 43, 58, 93–94; tinkerers and, 58, 65, 93; women and, 93

capitalism, racial, 12, 19, 110, 120, 149, 150, 181n8

capitalism: and affect, 157; and bioeconomies, 14–15; biology and, 14–15, 56; and Black people, 178n6; and categories of difference, 108; control of knowledge and, 46; co-optation of social justice concepts by, 8–9; and COVID-19, 171; and definition of failure, 125; and definition of success, 125; and democratic sciences, 15, 65; and disability, 140; and discrimination, 42; DIY biology and, 48–49, 62–63; as dominant over populace, 11; and the environmental crisis, 67; and exclusion, 140; and gender, 6–7; and genomic research, 14; globalization of, 8, 92; hacking and, 54; and humans as economic subjects, 64; and inclusion, 9, 21, 63, 64, 65, 69; innovation and, 63; as natural system, 108; and neoliberalism, 9, 159; and the open source movement, 67–68; and outsourcing labor, 67; and people of color, 65; play and, 62–63; and race, 6–7; and racism, 178n6; science and, 108; and scientific ethics, 153; scientific failures as challenge to, 121; and scientific success, 121; and social order, 56; success under, 130, 159; and systemic inequalities, 6, 125; in the United States, 54, 171, 178n6; and United States racial politics, 11–12; universities and colleges and, 8–9, 14, 32, 34; virtue signaling under, 42; women and, 6–7; and women's health, 141. *See also* entrepreneurship; innovation; power, economic; success, economic

capitalists, 12, 65, 94

Caporael, Linnda, 16

Chicana/o/x studies, 143

children, 63

Church, George, 35, 61
Clare, Eli, 99
Clarke, Jan, 142
class, social: and access to science, 83; of
  DIY biologists, 76; and nature (natural
  world), 111; scientists and, 55, 64–65; and
  social justice, 89; and success, 59;
  tinkerers and, 61–63; white supremacy
  and, 56
classes, DIY biology, 39, 45, 47, 79, 80, 165
collectivity, 131, 145, 149, 173, 174
colleges. *See* universities and colleges
coloniality: and categories of difference,
  156; concept of intelligence and, 56; and
  definition of the human, 121–122; and
  democracy, 89; and distribution of
  resources, 162; and DIY biology, 63; and
  economic power, 131; exploration and, 63;
  and gendered racial capitalism, 93; and
  inclusion, 108; and individualism, 131;
  knowledge and, 108; and lone scientists,
  130; and love, 156–157; and myth of
  objectivity, 120; rectification of, 154; and
  science, 7, 24, 69, 108, 124, 156–157; and
  social justice movements, 147; in STEM
  (science, technology, engineering, and
  mathematics), 100; and structural power,
  108, 121–122; and systemic inequalities,
  108; white men and, 56; white women
  and, 131; and women's health, 141
colorblindness: and definition of community, 132; and DIY biology communities,
  17, 69, 79; and inclusion, 79; and passion
  for science, 60–61, 110; and systemic
  inequalities, 43, 60; tinkerers and, 21,
  65, 68, 93, 94, 148; tinkering and, 16.
  See also race
Columbus, Christopher, 130
commercialism, 67
Committee on Women, Population, and
  the Environment (CWPE), 98, 145
Common Science? (Barr and Birke), 128, 142
communities, DIY biology: activists and,
  31t; and affect, 17; antiracism in, 49, 68; as
  apolitical, 165; and beneficial science,
  91–92; and biocitizenship, 78; claims of
  disenfranchisement by, 62; colorblindness in, 17, 69, 79; concern over image of,
  73–74; co-optation of social justice ideas
  by, 42, 62, 69, 92, 137; and COVID-19, 171;
  and creativity, 49; and definition of the
  human, 51; demographics of, 41–42, 50;
  as distinct from meth users, 76–77;
  education by, 28–29, 79; elitism in, 44;
  exclusion from, 41, 50; formation of, 17,
  28, 29, 74, 84; and gender, 41–42, 50–51,

65, 68; genderblindness in, 17, 69; and
  idea of scientific progress, 94–95;
  inclusion in, 41, 83; lack of diversity in,
  42, 43, 44, 50, 54, 65; lack of in majority
  Muslim countries, 89; and law enforcement, 77–78; and liberalism, 68; membership in, 31t, 32, 49; and passion for
  science, 51; perceived lack of racism in,
  44; politics of inclusion in, 20; as the
  public, 77, 78, 82, 83, 89–90; and race, 42,
  65, 68; as safe, 82; and safety regulations,
  75; and science literacy, 79, 81, 82; sexism
  in, 43, 44; and systemic inequalities, 41,
  68; and terrorism, 78; tinkerers and, 30;
  and traditional science, 78; as valuing
  diversity, 32, 68; white women in, 42;
  women in, 43–44, 51. *See also* laboratories, DIY biology; *specific DIY biology
  groups*
communities, DIY science, 79
communities, formation of, 165
computers, 61, 78
consent, 144
counterculture figures, 62, 77, 78, 82,
  89, 90
COVID-19, 3–4, 124, 170–171, 172
Cowell, Mac, 27
creativity, 11, 24, 49, 55, 57, 59, 67
critical ethnic studies, 19, 122
critical studies, 143
crowdfunding, 136
culture, 168

DARPA (Defense Advanced Research
  Projects Agency), 180n13
Darwin, Charles, 64, 129
Daston, Lorraine, 55
Dean, Jodi, 67
Death beyond Disavowal (Hong), 132
decoloniality, 19, 100–101, 156–157, 158,
  184n1
decolonial studies, 18, 89, 122, 143, 185n6
deconstruction, 185n3
defection from science, 21, 101–102, 110, 115,
  124, 183n1
Deleuze, Gilles, 144
Delfanti, Alessandro, 69
democracy: and coloniality, 89; and
  COVID-19, 170; direct, 171; feminist
  science and, 138; and neoliberalism, 11; and
  the public, 90, 148; rhetoric of, 93, 94; and
  science, 15, 21; in the United States, 171
demographics. *See under* communities,
  DIY biology
demos, the, 8, 22, 41, 64, 65, 86, 94. *See also*
  public, the

INDEX

Department of Defense, 78

Depo Diaries, 145

Depo-Provera, 145

Derrida, Jacques, 178n9, 185n3

difference, categories of, 132, 185n3; activists' use of, 146–147; and biological determinism, 65; and biology, 7; capitalism and, 108; and coloniality, 156; and concept of intelligence, 56; and empathy, 163; and feminist science studies, 127; and gendered racial capitalism, 140; as innate, 59; and intersectionality, 100; and love, 157, 158; and politics of inclusion, 150; and process (methodology), 162; and the public, 90; and reflexivity, 167–168; and reproductive justice networks, 153; and science, 68; scientists and, 58; and systemic inequalities, 65, 131

difference, politics of, 132

diffraction (metaphor), 167–168

disability: and ableism, 99; and access to science, 83; and capitalism, 140; and feminist studies, 127; and knowledge production, 149, 167; medical model of, 145, 146; and night politics, 152; science of measurement and, 56; and sitpoint theory, 184n2; social model of, 146; and systemic inequalities, 100

disability politics, 153

disability studies, 143

discrimination, 42, 108, 109, 111, 133, 147. *See also* inequalities, systemic

disenfranchisement: claimed by DIY biologists, 62; and gendered racial capitalism, 8, 131; and social justice, 89; tinkerers and, 21, 58, 61, 88, 90, 93, 94; by traditional science, 32; of white men, 11, 61

diversity: and knowledge production, 97; lack of in DIY biology communities, 42, 43, 44, 50, 54, 65; and passion for science, 60; in universities and colleges, 181n10; as valued by DIY biology communities, 32, 68

DIYbio (network), 28, 31–32, *33,* 75, 78, 79, 85, 182n16

DNA: ancestry markers in, 45; collection of, and consent, 144; labor in processing of, 14; and PCR (polymerase chain reaction) technology, 39; research on, and race, 49–50; use of in DIY biology, 3, 47, 48; use of in synthetic biology, 46, 85

DNA Dreams (documentary), 50

Doctors without Borders, 92

Dunbar-Hester, Christina, 67, 165

education: and affect, 60; Black Panther Party and, 149; and community needs, 81; cost of, 65; critical science, 128; by DIY biology communities, 28–29, 79; and feminist science, 124–125; feminist scientists and, 125; and gender, 40; promoted by tinkerers, 90; and queer theory, 122; and race, 111; and resistance to science, 115, 124; revaluing of, 133; roles for scientists in, 118; and systemic inequalities, 59, 84. *See also* pedagogy; teaching; universities and colleges

elitism, 44, 89

ELSI (ethical, legal, and social implications) programs, 84, 85

emotions, 7, 9, 39, 56, 107. *See also* affect; passion for science

empathy, 163

Endy, Drew, 45–46, 91

engineering, 16, 46–47, 57, 85, 178n9

engineers, 40, 46

entanglements, 168

entrepreneurs, 82

entrepreneurship: and beneficial science, 92; biohacking and, 54; Black people and, 13; and concept of innocence, 63; and democratic sciences, 151; DIY biologists and, 62–63, 82, 159; DIY biology and, 17, 29, 34, 47, 48–49, 54, 62–63, 68, 69; and exploration, 29, 34; and failure at science, 130; as genetically driven, 59; Genspace and, 63; and passion for science, 63–64; and play, 66; scientists and, 59; and systemic inequalities, 59; tinkerers and, 58, 64, 93; and tinkering, 13, 16, 49, 91, 151; white men and, 66. *See also* capitalism; innovation; success, economic

environmental crisis, 67

epistemic authority: academic scientists and, 2; access to, 17, 111; and feminist scientists, 130; and feminist studies, 183n4; and literacy, 123; and passion for science, 112; redistribution of, 118, 128, 133, 151; of science, 15, 60, 105, 108, 111, 112, 113, 130, 182n18, 183n6

epistemic power, 21, 22, 58, 60, 109, 112, 123, 133

epistemic privilege, 106

equality, 89

Espineira, Karine, 141

ethics, 145, 153, 167, 168, 185n3, 185n5

ethics, feminist, 151

ethics, scientific: and accountability, 86, 93; amateurs and, 74; of animal research, 153; and Big Bio, 81; biologists and, 85; and bioterror, 78–79; and capitalism,

ethics, scientific (cont.)
153; and democratic sciences, 93; DIY biologists and, 94; and DIY biology, 75–76, 85, 89, 94; and ELSI (ethical, legal, and social implications) programs, 84, 85; and fear of science, 75; and genetic engineering, 75, 81, 85; and genetic research, 21, 35, 85, 93; of genomic research, 93, 94; and innovation, 78; and knowledge production, 71, 90–91; and positionality, 93; and the public, 85, 86, 90, 93; and racism, 49, 50; scientists and, 86; and situated knowledge, 160; and synthetic biology, 35, 69, 74, 85; tinkerers and, 72, 90–91, 94; and tinkering, 17, 93; in universities and colleges, 70–71, 73
ethnic studies, 122, 183n4
ethnoscience, 7
eugenics, 55, 56, 147
Europe, 79
Evans, Antony, 34, 35
exclusion, politics of, 41
exclusion: affect and, 9; of Black people from science, 100, 110–111, 123; and capitalism, 140; and concept of intelligence, 59–60; from definition of the human, 18, 51; and democratic sciences, 6; and DIY biologists, 94; from DIY biology communities, 41, 50; feminist science studies and, 17; gender and, 30, 32, 41, 58–59; and gendered racial capitalism, 58, 69, 140; and identity formation, 17; and passion for science, 60–61; and positionality, 10; and the public, 86; race and, 30, 41, 58–59; and rationality, 57; from science, 17, 22, 32, 57, 115, 123; and social justice, 83; and systemic inequalities, 41, 65, 100; tinkerers and, 21, 58, 61; tinkering and, 16
exploration, 29, 34, 39, 46, 49, 63, 82

facultad, la (concept), 163–164
failure, 121, 122, 125, 129, 130, 134
families, 7, 9, 107, 114t, 184n6
Fausto-Sterling, Anne, 56, 98, 109, 111, 113, 115, 127
Federal Bureau of Investigation (FBI), 27–28, 76–78
Feeling for the Organism, A (Keller), 67
femininity, 46–47
feminism: challenges to racial capitalism by, 149; challenges to structural power by, 105; challenges to systemic inequalities by, 100–101; and decoloniality, 100–101; and democratic sciences, 91, 142; and failure, 122; and gendered racial capital-

ism, 140; and intersectionality, 184n7; and knowledge production, 17, 160; and passion for science, 109; and pedagogy, 127; and play, 152; and race, 149; and racism, 132; vs. science, 102, 103, 104–105, 106, 108, 111–112, 117–118; and science literacy, 82, 126; and tinkering, 16, 142, 144–145, 162
feminist killjoys, 82, 107, 113
feminist movements, 32, 62, 116, 138
feminist politicization, 163
feminists, 10, 149, 184n7. *See also* scientists, feminist
feminist science studies: and categories of difference, 127; decoloniality and, 19; definition of, 101; and democratic sciences, 6; and ethics, 168; and exclusion, 17; feminist science and, 101; and feminist studies, 18, 19; and inclusion, 17; and knowledge production, 125, 151; and laboratories, 133; liminality of, 103; misunderstanding of, 42; and passion for science, 109; and pedagogy, 125; and queer theory, 122; and racialization, 100; right-wing responses to, 120; and standpoint theories, 19; and structural power, 86; and systems of oppression, 17–18; in universities and colleges, 109; and women's studies, 102–103. *See also* science, feminist
Feminist Science Studies: A New Generation (Mayberry, Subramaniam, and Weasel), 101, 102–103, 125, 142
feminist spies, 124, 184n3
feminist studies: as academic discipline, 105, 127; and disability, 127; and epistemic authority, 183n4; and feminist science studies, 18, 19; feminist scientists and, 99, 126; and knowledge production, 124–125, 126; as maintaining systemic inequalities, 105; and passion for science, 107; positionality of, 105; and race, 183n5; and science, 21, 102; and science literacy, 113, 127. *See also* women's studies
feminization, 40, 44, 50, 66, 170
Feynman, Richard, 82
Foege, Alec, 11, 13
Foster, Ellen, 165
Foucault, Michel, 15, 108
funding for science, 87t; Charles Darwin and, 64; and crowdfunding, 136; and DARPA (Defense Advanced Research Projects Agency), 180n13; and disability research, 99; and DIY biology, 34, 42, 48, 61, 83; feminist science and, 104; by the government, 88; and inclusion, 17; as

INDEX

institutional privilege, 174; and politics of inclusion, 42; and the public, 85; tinkerers and, 62; and traditional science, 2, 70, 82, 104; in universities and colleges, 183n3

Galison, Peter, 55
Garland-Thomson, Rosemarie, 184n2
gatekeeping, 29, 32, 83–84, 96
geeks, 31t; computer, 31, 54, 67; as male, 44, 65; and masculinity, 44, 65; and neoliberalism, 67–68; and the open source movement, 67; and passion for science, 59; as sexist, 43; and tinkerers, 65; as white men, 65
gender: and access to science, 83; as applied to exploration, 46; bioeconomies and, 15; biohackers and, 138; and capitalism, 6–7; and concept of intelligence, 56, 60; and definition of the human, 7, 11; DIY biologists and, 65–66; and DIY biology communities, 41–42, 50–51, 65, 68; and DIY biology labs, 17; and economic power, 65; and education, 40; in engineering, 46–47; and exclusion, 30, 32, 41, 58–59; and genetic research, 65; hierarchies of, and science, 5; illiteracy toward, 123; and labor, 42, 65, 66, 170; and nerd culture, 40; and night politics, 152; and passion for science, 41, 65; as performative, 167; as racialized category, 100; and rationality, 56; and science, 177n5; science of measurement and, 56; and systemic inequalities, 116–117; tinkerers and, 43, 59; tinkering and, 69; and transfeminism, 140–141; and Trumpism, 12. See also men; sexism; women
genderblindness: and definition of community, 132; in DIY biology communities, 17, 69; and passion for science, 60–61; and systemic inequalities, 43, 60; tinkerers and, 21, 65, 68, 93, 94; and tinkering, 16. See also sexism
gender essentialism, 177n5
genderf-cking, 167
genderhacking, 167
gender roles, 40, 42
genealogies (historical technique), 22, 108, 138, 139–140, 162
genetically modified organisms (GMOs), 35
genetic engineering: activists and, 35; and fear of science, 85; and scientific ethics, 75, 81, 85; and synthetic biology, 27, 45–46, 85; viewed as feminized, 46–47, 50–51
genetic research: automation of, 46, 47; and consent, 144; criticism of, 35; DIY biology classes in, 39; in DIY biology laboratories,

48; and gender, 65; labor involved in, 45, 47, 66; and the public, 21; and scientific ethics, 21, 35, 85, 93; and synthetic biology, 2–3, 46–47; and tinkering, 16
genetics (of a person), 16, 48, 49–50, 59. *See also* biological determinism
genetic sequencing, 46
Genome Project, 84
genomic medicine, 123
genomic research, 14, 15–16, 45, 84–85, 93, 94, 123
Genspace, 31t, 36, 37, 38; and Biotech without Borders, 92; classes offered by, 39, 79, 171; and democratic sciences, 32; and entrepreneurship, 63; founding of, 28; and gentrification, 25; mission of, 29; neighborhood environment of, 37; and property rights, 48; and the public, 80, 90; and safety, 82; and tinkering, 59
gentrification, 25–26
Gentry, Eri, 58
Gesturing Towards Decolonial Futures (collective), 101
Getting Lost (Lather), 156
Gill, Rosalind, 43
Gilmore, Ruth, 170
globalization, 8, 12, 92
Goldman, Emma, 152
Google (company), 66, 181n8
Gould, Stephen Jay, 55
Gschmeidler, Brigette, 45
Guattari, Félix, 144
Gynepunk, 141

hackers: computer, 68; as counterculture figures, 89; definition of, 31t, 180n13; DIY biologists as, 76; identities of, 138; and military funding, 180n2; positionality of, 77; vs. terrorists, 76; work ethic of, 54
hackerspaces, 54, 138
hacking, 54, 167
Halberstam, Jack, 121, 125, 129–130
Hammonds, Evelynn, 55, 110–111
Hamraie, Aimi, 143, 146, 162–163, 167
hands-on learning, 162, 165–166
Haraway, Donna, 19, 145, 156, 161, 167–168, 169, 185nn3–4
Harding, Sandra, 19, 115, 152
Harris, Dennis, 16
health care, 139, 147, 150, 154–155. *See also* activists, health; medicine; women's health movement
Hemmings, Clare, 117, 162, 163
Hess, Amanda, 177n4
Hessel, Andrew, 136, 137
heteronormativity, 7, 9, 12, 65, 122

heteropatriarchy, 7
heterosexuality, 12, 40, 181n5
HIV/AIDS movements, 56, 139, 150
Homeland Security, 76–78
Homo economicus, 15, 64, 65, 66, 150
homophobia, 11, 44
Hong, Grace, 132, 185n3
Hubbard, Ruth, 111, 115
human, definition of the, 7, 11, 18, 51, 108, 121, 149
human beings, 8, 64, 160
Human Genome Research Institute, 84
humanities, the, 56, 103, 104, 162
humanity, 156–157

identity, 16, 17, 132, 140, 148, 160, 180n14
identity politics, 11, 69, 141, 149, 150
inclusion, politics of: Black people and, 132; and categories of difference, 150; and DIY biology, 30, 94, 140; in DIY biology communities, 20; feminist science and, 122–123; and funding for science, 42; and gendered racial capitalism, 140; and justice, 89; liberal, 140; and people of color, 30; and structural power, 181n10; tinkerers and, 65; in universities and colleges, 8, 181n10; and the women's health movement, 150
inclusion: and capitalism, 9, 21, 63, 64, 65, 69; and coloniality, 108; colorblindness and, 79; and definition of the human, 51; and the demos, 86; DIY biologists and, 89; in DIY biology communities, 41, 83; economic success as goal of, 58; and feminist science studies, 17; and funding for science, 17; and gatekeeping, 84; and gendered racial capitalism, 140; and liberalism, 140; and passion for science, 17, 60, 65, 69, 111; positionality and, 10; and the public, 66; and racialization, 108; rhetoric of, 93; in science, 17, 22; situated nature of, 89; and social justice movements, 11; and systemic inequalities, 100, 132, 133; tinkerers and, 90, 148; of women, 139–140
Indigenous knowledge, 185n4
Indigenous people, 5, 91, 93, 100, 171
Indigenous studies, 18, 143
individualism, 131, 138, 140, 149, 159, 168, 173
inequalities, systemic: and access to science, 10; and biological determinism, 9; and capitalism, 6, 125; and categories of difference, 65, 131; and coloniality, 108; and colorblindness, 43, 60; and co-optation of social justice ideas, 172; and COVID-19, 170–171; and creation of

laboratories, 133; and democratic sciences, 89; and disability, 100; and DIY biology communities, 41, 68; and education, 59, 84; and entrepreneurship, 59; and exclusion, 41, 65, 100; feminist challenges to, 100–101; and feminist science, 118; and gender, 116–117; and genderblindness, 43, 60; and gendered racial capitalism, 8, 58, 93–94; and identity, 160; and inclusion, 100, 132, 133; and labor, 132, 170; in laboratories, 14; as maintained by bioscientific knowledge, 15; as maintained by feminist studies, 105; and nature (natural world), 111; and neoliberalism, 41; passion for science and, 60–61, 66, 67, 110–111; play as rebellion against, 159; and positionality, 93; and poverty, 149; and race, 100, 116; rectification of, 154; science and, 7, 68–69, 108, 116–117, 154–155; and science illiteracy, 115; and science literacy, 122; structural power and, 111; in universities and colleges, 159–160, 172–173. *See also* discrimination; racism
informed consent, 73, 115, 145
informed refusal, 115
innocence, concept of, 63, 66, 69
innovation: and capitalism, 63; DIY biology and, 17, 32, 49, 58–59; passion for science and, 61; and science, 63; and scientific ethics, 78; tinkerers and, 62, 94; tinkering and, 10, 57. *See also* capitalism; entrepreneurship
institutional review boards (IRB), 73
intellectual property, 14, 67
intelligence, concept of, 49, 56, 59–60, 65, 69, 105, 106, 107
intelligence (quality), 9, 20–21, 30, 56, 110
interdependence, 149
interdisciplinarity, 103–104
intersectionality, 7, 100, 141, 149, 157–158, 171, 184n7
intersex, 178n5
intra-action, 168
Irani, Lilly, 64
Iraq, 89
iteration, 166, 167, 168–169, 173, 174

Jasanoff, Sheila, 88
Jen, Clare, 65
Jobs, Steve, 61, 78, 138, 181n4
Jorgensen, Ellen, 58, 75, 76
justice, 89, 185n4
justice, disability, 141
justice, environmental, 5, 138, 139, 141, 145
justice, health, 141

INDEX

justice, racial, 62

justice, reproductive, 141, 145, 149, 152–153, 184n1

justice, social: and coloniality, 147; co-optation of ideas from, 172; as co-opted by capitalism, 8–9; as co-opted by DIY biologists, 93; as co-opted by DIY biology communities, 42, 62, 69, 92, 137; as co-opted by neoliberalism, 11; as co-opted by tinkerers, 88, 90, 93, 95; and democratic sciences, 88, 91; and disenfranchisement, 89; and exclusion, 83; funding for, 174; and health care, 147; and inclusion, 11; and knowledge production, 86, 88, 93; as knowledge production, 143; and neoliberalism, 93; and pedagogy, 127, 128; and political ideology, 139; rights discourse within, 83; and science, 15, 98, 165; and science literacy, 81; and social class, 89; as tinkering, 147; and virtue signaling, 42

Kazi, Nazia, 12

Keller, Evelyn Fox, 111, 116–117, 125, 131, 184n7

Kelty, Christopher, 62, 85, 89

Kickstarter, 34, 35

Kirkpatrick, Graeme, 54

knowledge, control of: biohackers and, 90; and biological sciences, 56; and the Black Panther Party, 148; and capitalism, 46; and democratic sciences, 81; and DIY biology, 46, 67; and nature (natural world), 8; and poverty, 149; by scientists, 56; and self-determination, 147; structural power and, 149; and synthetic biology, 46; and the women's health movement, 148

knowledge: bioscientific, and systemic inequalities, 15; and coloniality, 108; critical science literacy and, 172; Indigenous, 185n4; and *la facultad*, 163–164; and medical noncompliance, 115; piracy of in biology labs, 14; as praxis, 168; racialization of, 108; right to, 84; as situated, 86, 88, 145–146, 151, 156, 160, 161, 184n1; Sylvia Wynter and, 143

knowledge economies, 14

knowledge production: and accountability, 18, 151, 160–161, 168–169; by activists, 143, 162–163; and affect, 164; antiscience approaches to, 124; under capitalism, 14; challenges to by activists, 14; as contingent, 112; control of, 8, 15, 17, 123; and creativity, 57; critical science literacy as, 143, 146; critiques of science as, 183n6; cultural assumptions and, 99; Deboleena

Roy and, 126; decolonization of, 19; and disability, 149, 167; by disability activists, 143, 149, 167; and diversity, 97; and DIY biology, 34, 47; by DIY biology labs, 34, 50; and feminism, 160; feminist laboratories and, 99–100; and feminist science studies, 125, 151; feminist studies and, 124–125, 126; and gendered racial capitalism, 58, 65; health activism as, 143; and inequality, 15; and interdisciplinarity, 103–104; lab techs and, 130–131; and love, 155; by marginalized people, 100; narratives as, 145; by *oncogrrrls*, 167; and passion for science, 66; pedagogy as, 127; performance as, 167; and play, 155–156, 161–162; process of, 162, 164–165; by the public, 81; and reassembly, 166; and responsibility, 143; and science, 60, 120; and science literacy, 128; and scientific ethics, 71, 90–91; scientific failures as, 121; and self-determination, 147; as situated, 161, 168; social impacts of, 18; and social justice, 86, 88, 93, 143; and standpoint theories, 145; and structural power, 86, 127; subjects of, 123; teaching as, 127; tinkerers and, 17, 57; tinkering and, 53; and transfeminism, 165–166; and truth, 15, 55, 99, 105; and the women's health movement, 144, 145, 148

LA Biohackers, 25, 29, 30, 31t, 37, 76

"Lab of Her Own, A" (Anft), 10

"Lab of Their Own, A" (Kean), 9–10, 61

labor: and citizenship, 8; DIY biology as, 45; feminist scientists and, 130–131; and gender, 66, 170; gendered division of, 42, 65; grad students as, 45; involved in genetic research, 45, 47, 66; and laboratories, 184n5; of lab techs, 14, 47, 130–131; outsourcing of, 66, 67; and people of color, 125; and race, 66; racialization of, 65; and systemic inequalities, 132, 170; theft of under capitalism, 7

laboratories, DIY biology, 37, 38; activists and, 25; as distinct from meth labs, 76–77; and gender, 65; genetic research in, 48; lack of equipment in, 48; and positionality of DIY biologists, 76; public access to, 29, 32; and safety, 75, 76, 78, 82; and tinkerers, 13; vs. traditional laboratories, 34, 67; in the US, 79. *See also* communities, DIY biology; *specific DIY biology groups*

laboratories, traditional, 17, 32, 34, 59, 67, 78. *See also* Big Bio; universities and colleges laboratories: creation of, and systemic inequalities, 133; definition of, 126; feminist, 99–100, 126–127; feminist science studies and, 133; and labor, 184n5; outsourcing of labor in, 66, 67; and pedagogy, 126; structural power in, 130; systemic inequalities in, 14

"Laboratories of Our Own: New Productions of Gender and Science" (Subramaniam), 10

laboratory materials, 130

La Paillasse (DIY biology lab), 30, 31t

Lather, Patti, 153, 155, 156, 160–161, 166

Latin America, 79

Latinx people, 100, 154

Latour, Bruno, 131

law enforcement, 27–28, 76–78

Leninism, 147

Lévi-Strauss, Claude, 178n9

liberalism, 68, 138, 140, 177n4

liberation, 159

liberation movements, 11, 139

liminality, 103, 132–133

literacy, 123

Lorde, Audre, 132, 139, 145, 184n7

love, 155, 156–158

Lugones, María, 17, 155, 156, 157–159, 160, 161, 162, 169

Maher, Frances, 127

man, bioeconomic, 15

Maoism, 147

March for Science (2017), 4, 5

Martin, Emily, 115

Martin, Paul, 85

Marxism, 146, 147

masculinism, 16, 51, 150, 152

masculinity, 152; geeky, 44, 65; nerd, 96, 106; and passion for science, 109; racialized perceptions of, 40; and tinkerers, 69; white, 65, 69, 109

materialism, 100, 105, 162

materialism, new, 162

material resources, 162, 173–174

Maxigas, 138

Mayberry, Maralee, 127, 142

McClintock, Barbara, 67, 161

McIntosh, Peggy, 105

measurement, science of, 55–56

medical apartheid, 123

medical authority, 115, 123, 149

medical civil rights movement, 139

medicine, 8, 80, 147, 166. *See also* activists, health; Black Panther Party; health care

men, 43, 44, 177n5

men, Asian, 40

men, Black, 40

men, white: and capitalism, 7; as capitalists, 94; and coloniality, 56; and concept of intelligence, 56; as disenfranchised, 11, 61; and DIY biology demographics, 42; emotions of as rational, 56; as engineers, 40; and entrepreneurship, 66; geeks as, 65; as human, 11; and meth labs, 76, 77; and nerd culture, 96; and objectivity, 55; and passion for science, 65; and rationality, 56; as respectable, 78; science of measurement and, 56; as scientists, 39, 40, 55, 56, 58, 59, 129; and social class hierarchy, 56; as tinkerers, 12, 64; United States conservative, 11–12. *See also* gender

meritocracy, 159–160

meth labs, 76–77

methodologies, 151–152

Meyer, Morgan, 89, 90

Mignolo, Walter, 184n1

modernity, 100–101

Mohamed, Ahmed, 12

Mullis, Kary, 39–40

Murphy, Michelle, 139, 144, 149, 166

Musil, Caryn McTighe, 127

Muslims, 79, 89

mutual aid, 171

nanotechnology, 91

Narek, Diane, 116

narratives, 61, 69, 145

National Academies of Science, Engineering, and Medicine, 75

National Institutes of Health (NIH), 70

National Organization for Women, 139

National Science Foundation (NSF), 127

nature (natural world): and affect, 109; control of knowledge and, 8; and myth of objectivity, 55, 55t, 87t; and science, 105, 108, 113, 183n7, 184n7; and situated knowledge, 151; and synthetic biology, 46; and systemic inequalities, 111; tinkerers and, 57, 64; and tinkering, 16

naturecultures, 168

negative results, 121

Nelson, Alondra, 139, 147, 148

neoliberalism: and antiracism, 11, 139; and biohacking, 138; and capitalism, 9, 159; as co-opting social justice ideas, 11, 93, 140, 157; and democracy, 11; and democratic sciences, 88; and economic power, 11; and embrace of failure, 121; geeks and, 67–68; and medicine, 166; and the open source movement, 68; and play, 157; racism and backlash against, 11; and systemic inequalities, 41; and tinkering, 16, 151–152; and traditional science, 90; and the women's health movement, 139, 166

INDEX

nerd culture, 40, 43, 59, 65, 96, 106, 109
neutrality, myth of, 67–68, 95, 142
"Never Meant to Survive: A Black Woman's Journey" (Hammonds), 110–111
Newton, Huey P., 147
New View of a Woman's Body, A (Federation of Feminist Women's Health Centers), 145, 146
night politics, 140, 152
noncompliance, medical, 115, 123
normality, concept of, 71–72, 121
"Notes on a Transfeminist Technology" (Rojas), 123
Novella, Caro, 167

Obama, Barack, 85
objectivity, myth of, 24–25, 55, 55t, 56, 87t, 120, 142, 156
objectivity: and concept of innocence, 63; feminist, 142, 151; as quality for scientists, 39; science and, 7; as social construct, 54–55, 55t; social ineptitude as sign of, 41; and standpoint theories, 56, 156; and tinkering, 57–58; and white men, 55
oncogrrrls, 152, 166, 167
Open Insulin Project, 171
open source movement, 48, 67–68
Open Wetlab, 79
oppression, systems of, 17–18. *See also* inequalities, systemic
Our Bodies, Ourselves (Boston Women's Health Book Collective), 8, 146, 148

Panichkul, E. Gabriella, 16
passion for science, 87t; biologists and, 39–40; colorblindness and, 60–61, 110; and democratic sciences, 57; and discrimination, 111; and diversity, 60; DIY biologists and, 30; DIY biology and, 58–59, 69; in DIY biology communities, 51; and entrepreneurship, 63–64; and epistemic authority, 112; as exclusionary, 60–61; exploration as part of, 39; and feminism, 109; and feminist science studies, 109; feminist scientists and, 107, 109, 134–135; feminist studies and, 107; and geeks, 59; and gender, 41, 65; genderblindness and, 60–61; as heteronormative, 65; and inclusion, 17, 60, 65, 69, 111; and innovation, 61; intelligence and, 9, 20–21, 30, 110; and knowledge production, 66; masculinity and, 109; the public and, 66; and race, 65, 111; Richard Feynman's, 82; scientists and, 40–41, 56, 59; and systemic inequalities, 60–61, 66, 67, 110–111; tinkerers and, 63–64, 69; and truth, 114t; underdog narratives and,

61–62; and white men, 65; women and, 41, 110. *See also* affect; emotions
Patterson, Meredith, 83
PCR (polymerase chain reaction) technology, 39, 41
pedagogy, 125–129, 153–154, 162, 164, 167, 172–173. *See also* education; teaching
people of color: and biological determinism, 147; and capitalism, 65; and concept of intelligence, 56; and COVID-19, 170; as excluded from science, 32; and gendered racial capitalism, 93; health activism and, 139; and labor, 125; and politics of inclusion, 30; and the public, 87; in science, 97, 122, 123; at universities and colleges, 8. *See also* Black people; race
People's Community Medics (PCM), 164
performance, 167
performativity, 167
Pfizer, 49
physics, 82
Pilgrimages/Peregrinajes (Lugones), 159, 161
Pink Army Cooperative, 136, 137
pinkwashing, 137
play (activity): animal research about, 71; and biohacking, 155; and capitalism, 62–63; as co-creation, 156; and community building, 169; as co-opted by neoliberalism, 157; DIY biology as, 29, 44–45, 62–63, 82, 83; and entrepreneurship, 66; and feminism, 152; feminist science as, 160–161; feminist tinkerers and, 162; and knowledge production, 155–156, 161–162; and liberation, 159; and love, 155, 158; as rebellion against systemic inequalities, 159; science as, 66, 155; synthetic biology as, 45; tinkerers and, 158–159; tinkering as, 16–17, 155; traditional science as being without, 84; and transfeminism, 153
politics, environmental, 5
politics, United States, 11–12
populism, 11, 12, 15
positionality: and activists, 156; of biohackers, 78; of DIY biologists, 76–77, 83; and exclusion, 10; of feminist studies, 105; of hackers, 77; and identity, 132; and inclusion, 10; of punks, 78; and scientific ethics, 93; of scientists, 24–25, 86; and systemic inequalities, 93; of terrorists, 78; of tinkerers, 10, 24–25, 53, 61–63, 64–65, 93; and tinkering, 165; and truth, 24, 25; of white people, 42
postcolonial studies, 89, 122
postfeminism, 11, 68, 69, 84, 89
postfeminist sensibility, 43, 44
posthumanism, 66

postmodernism, 120, 160, 162
post-porn movements, 140–141
postracialism, 11, 68, 69, 78, 84, 89
poststructuralism, 160
poverty, 12, 18, 25, 65, 78, 147, 149, 166
power, analysis of, 150
power, economic: activists and, 139; and coloniality, 131; and COVID-19, 172; and defection from science, 125; and democratic sciences, 174; and gender, 65; and gendered racial capitalism, 58, 93; and neoliberalism, 11; and race, 65; shifts in, 8; and tinkering, 13, 63–64, 68, 69, 91
power, modes of, 146
power, personal, 152
power, political, 8–9, 15, 147
power, politics of, 106, 130
power, racialized, 183n5
power, structural: access to through science, 111; and accountability, 133; and anticapitalism, 121; of biology, 58; building of, 150; and coloniality, 108, 121–122; and control of knowledge, 149; and democratic sciences, 88; feminist challenges to, 105; feminist scientists and, 106; and knowledge production, 86, 127; in laboratories, 130; and liminality, 103; loss of, 125; obscuring of, 151; and politics of inclusion, 181n10; reconfiguration of, 132; of science, 104, 110, 111–112, 125, 135, 151; scientists and, 105, 129; and systemic inequalities, 111; tinkerers and, 93
praxis, 161, 168
Presidential Commission for the Study of Bioethical Issues, 85, 94
privilege, 105–106, 131, 173–174
process (methodology), 162, 164–166
progress, 10, 47, 63, 84, 91, 94–95, 101
property rights, 48–49
public, the, 81t; and categories of difference, 90; and COVID-19, 171–172; definition of, 80, 86–87, 88, 90; and democracy, 90, 148; and democratic sciences, 21, 22, 88, 92; DIY biologists as, 86, 89, 92; and DIY biology, 75, 85; DIY biology communities as, 77, 78, 82, 83, 89–90; exclusion and, 86; and funding for science, 85; and gendered racial capitalism, 86, 91; and genetic research, 21; and Genspace, 80, 90; and inclusion, 66; and medicine, 80; and the myth of objectivity, 87t; and passion for science, 66; and people of color, 87; and science, 81, 82, 87–88; and science literacy, 2, 9, 79, 86, 88, 90; and scientific ethics, 85, 86, 90, 93; social scientists as, 86; and synthetic biology, 85; tinkerers as, 90; and

tinkering, 12, 13–14; and traditional science, 32, 57; women and, 87. *See also* demos, the
public relations, 28–30, 79–80, 81, 82
punks, 78

Queer Art of Failure, The (Halberstam), 121, 125, 129–130
queerness, 121, 135
queer politics, 153
queer studies, 18
queer theory, 122, 141
Quijano, Aníbal, 108
Quimera Rosa, 140–141, 152, 153, 157, 165–166, 167, 168

race: and access to science, 83; bioeconomies and, 15; and biological determinism, 147; and capitalism, 6–7; and concept of innocence, 63; and concept of intelligence, 56, 59–60; and definition of the human, 7, 11, 18, 108; and DIY biology communities, 42, 65, 68; and DIY biology labs, 17; DNA research and, 49–50; and economic power, 65; and education, 111; and exclusion, 30, 41, 58–59; and feminism, 149; and feminist studies, 183n5; hierarchies of, 5, 108; and labor, 66; negative effects of science and, 19; and nerd culture, 40; and passion for science, 65, 111; science of measurement and, 55–56; and systemic inequalities, 100, 116; tinkerers and, 59; tinkering and, 69; in United States politics, 11–12; and the women's health movement, 149. *See also* colorblindness; people of color
racialization: and DIY biology, 92; and feminist science studies, 100; and gender, 100; and inclusion, 108; of knowledge, 108; of labor, 65; of punks, 78; of terrorists, 76, 78–79; of tinkerers, 12, 93
racism, scientific, 5, 8, 49–50, 55–56. *See also* biological determinism
racism: and backlash against neoliberalism, 11; and capitalism, 178n6; definition of, 170; environmental, 5, 139; and family structure, 9; and feminism, 132; and love, 158; opposed by anticapitalists, 11; perceived lack of in DIY biology communities, 44; and science, 141, 146–147; and scientific ethics, 49, 50; in the United States, 170
rationality, 7, 9, 39, 56, 57, 64
Reardon, Jenny, 94
reassembly (concept), 144, 147, 166–167
Rees, Margaret, 127

Reeve, Christopher, 99
Reflections on Gender and Science (Keller), 131
reflexivity, 86, 167–168
religion, 87, 87t
research, animal, 70–71, 73
research, human, 73
resistance, politics of, 123
resistance to science, 21, 113–115, 124, 126–127, 128, 183n8. *See also* antiscience positions; science, fear of
response-ability, 169
responsibility, 84, 143, 157, 168, 185n3. *See also* accountability
revolution, 31–32, 62, 74, 78, 80, 136, 140
rights discourse, 84
Riley, Donna, 40
Rojas, Lucía Egaña, 123
Room of One's Own, A (Woolf), 10
Roosth, Sophia, 91
Rose, Hilary, 130, 131
Roy, Deboleena: and critical science literacy, 129; and feminist science, 117–118, 130, 151, 156, 160–161, 165; and feminist spies, 184n3; and iteration, 166; and knowledge production, 126; and Patti Lather, 153; and the reproductive justice movement, 142

safety, 75, 76, 78, 82, 103
Sandoval, Chela, 6, 157, 158, 164
Sands, Aimee, 110
San Francisco Bay Area, 25
Schiebinger, Londa, 56
science, democratic: and structural power, 88
science, fear of, 75, 78, 80, 81t, 82, 85, 112–115, 125. *See also* antiscience positions; resistance to science
science, feminist: and accountability, 132, 154–155, 166; and antiracism, 148–149; Deboleena Roy and, 117–118, 130, 151, 156, 160–161, 165; and democracy, 138; and democratic sciences, 98, 142; and education, 124–125; and feminist science studies, 101; and funding for science, 104; and methodologies, 152; as play, 160–161; and politics of inclusion, 122–123; and situated knowledge, 160; and systemic inequalities, 118; universities and colleges and, 102, 131; and women's studies, 113
science, rehabilitation, 146
science, traditional: activists and, 32; contextualization of, 130; defection from science and, 115; vs. democratic sciences, 90, 94; disenfranchisement by, 32; DIY

biologists and, 82, 83, 90, 92; vs. DIY biology, 78, 80, 89; DIY biology communities and, 78; feminist scientists and, 102; and funding for science, 2, 70, 82, 104; irreverence toward, 133; and neoliberalism, 90; as not playful, 84; and the public, 32, 57; vs. tinkering, 148. *See also* Big Bio; biology; laboratories, traditional; sciences, academic
science: accessibility of, and women's studies, 129; and accountability, 81, 82, 133; activists and, 124; and affect, 15, 60, 108, 113, 114t, 156, 172; antiscience positions and, 100, 113, 120, 123, 124; as beneficial, 91–92; and capitalism, 108; and categories of difference, 68; and coloniality, 7, 24, 69, 108, 124, 156–157; critical science literacy as, 127–128; critiques of, 183n6; and decoloniality, 184n1; definition of, 182n18; and democracy, 15, 21; as discipline, vs. tinkering, 30; and discrimination, 108, 109, 133; epistemic authority of, 15, 60, 105, 108, 111, 112, 113, 130, 182n18, 183n6; epistemic power of, 21, 60, 109, 112, 133; exclusion from, 17, 22, 32, 57, 115, 123; vs. feminism, 102, 103, 104–105, 106, 108, 111–112, 117–118; and feminist studies, 21, 102; feminist tinkerers and, 156; and gender, 177n5; and gendered racial capitalism, 68–69; and gender hierarchies, 5; impact of, and science literacy, 131; inclusion in, 17, 22; and innovation, 63; and knowledge production, 60, 120; and myth of neutrality, 95; and nature (natural world), 105, 108, 113, 183n7, 184n7; objectivity and, 7; people of color in, 97, 122, 123; as play, 66, 155; and political orientation, 3–4; and political power, 15; and the public, 81, 82, 87–88; race and negative effects of, 19; and racial hierarchies, 5; and racism, 141, 146–147; resistance to, 21, 113–115, 124, 126–127, 128, 183n8; as situated, 168; slogans concerning, 4, 4; and social justice, 15, 98, 165; structural power of, 104, 110, 111–112, 125, 135, 151; Sylvia Wynter and, 51, 108, 121–122, 150; and systemic inequalities, 7, 68–69, 108, 116–117, 154–155; and truth, 56, 100, 112, 177n5, 182n18, 183n6; and truth, absolute, 7; white women in, 100, 123; women in, 97, 99–100, 108, 116, 122, 123; and the women's health movement, 56; and women's studies, 99, 101, 109, 125, 127. *See also* funding for science; laboratories; passion for science; *specific scientific disciplines*

science appreciation, 147
Science for the People, 172, 185n1
science illiteracy, 115, 123, 124
science literacy, critical: and feminist scientists, 129; feminist tinkering as, 172; and knowledge, 172; as knowledge production, 143, 146; as science, 127–128; teaching of, 173
science literacy: and affect, 14, 82, 147; and COVID-19, 171–172; and DIY biology, 80; and DIY biology communities, 79, 81, 82; and feminism, 82, 126; feminist scientists and, 106; and feminist studies, 113, 127; and impact of science, 131; and knowledge production, 128; and the public, 2, 9, 79, 81, 86, 88, 90; purposes of, 2; resistance to, 113; and resistance to science, 115; as social justice issue, 81; and systemic inequalities, 122; and women's studies, 113
sciences, academic, 57, 59–60, 67, 89
sciences, democratic: and accountability, 82; activists and, 14, 139; as beneficial, 91–92; vs. Big Bio, 94–95; within bioeconomies, 14–15; and capitalism, 15, 65; and control of knowledge, 81; definition of, 3; and economic power, 174; emotional approach to, 9; and entrepreneurship, 151; and exclusion, 6; and feminism, 91, 142; and feminist science, 98; and feminist science studies, 6; and gendered racial capitalism, 89, 91; Genspace and, 32; and the "hard" sciences, 2; and neoliberalism, 88; new, 138; and opposition to gatekeeping, 32; and passion for science, 57; political aspects of, 5; and positionality of scientists, 25; and the public, 21, 22, 88, 92; and scientific ethics, 93; social and political stakes of, 18; and social justice, 88, 91; and systemic inequalities, 89; and tinkerers, 10; vs. traditional science, 90, 94; transitions in ethos of, 66; and Trump presidency (2016), 12. See also biology, DIY; biology, synthetic
"Scientific Literacy → Agential Literacy = (Learning and Doing) Science Responsibly" (Barad), 128
scientific method, 30
scientists, academic, 2, 14
scientists, feminist: as activists, 142; and affect, 109; and biohacking, 138; and critical science literacy, 129; and defection from science, 21, 101–102, 110; and education, 125; and epistemic authority, 130; and feminist studies, 99, 126; and interdisciplinarity, 104; and labor issues, 130–131; lack of, in synthetic biology, 137; and passion for science, 107, 109, 134–135; and politics of inclusion, 122–123; and science literacy, 106; and scientific subjects, 161, 165; and structural power, 106; and traditional science, 102; and the women's liberation movement, 116; and women's studies, 103–104, 106, 117
scientists: and categories of difference, 58; and the concept of intelligence, 56, 105, 106, 107; and the concept of normality, 71–72; control of knowledge by, 56; and creativity, 55; definition of, 65, 87, 87t; and entrepreneurship, 59; epistemic power of, 58; lone, 129, 130; male, as partners to women, 40; and objectivity, 39; and passion for science, 40–41, 56, 59; positionality of, 24–25, 86; roles for in education, 118; and scientific ethics, 86; and social class, 55, 64–65; and structural power, 105, 129; tinkerers as, 57, 58–59; training in ethics of, 70; and truth, 57; white men as, 39, 40, 55, 56, 58, 59, 129; and women's studies, 106; work done by, 184n4
Seale, Bobby, 147
Seiringer, Alexandra, 45
self-determination, 7, 8, 147, 148, 149, 164
Self-Help Clinic, 144
self-reflection, 155, 159, 160, 161
serendipity, 24
Sexing the Body (Fausto-Sterling), 98
sexism, 4, 11, 43, 44. See also gender; genderblindness; women
sexuality, 56, 141, 152–153
Sims, J. Marion, 141
sitpoint theory, 184n2
slogans, political, 4–5, 177n4
Smith, David, 67
social media, 80, 124
social sciences, 104
social scientists, 85–86
South Africa, 124
Spanier, Bonnie, 101, 103, 115–116
Spivak, Gayatri, 185n3
Stamboliyska, Rayna, 77–78
standpoint theories: activists and, 146; and feminist science studies, 19; and knowledge production, 145; and la facultad, 163, 164; and objectivity, 56, 156; and tinkering, 16; and the women's health movement, 145–146. See also sitpoint theory
STEM (science, technology, engineering, and mathematics), 42, 60, 100, 104
subjects, scientific, 161

Subramaniam, Banu, 10, 59, 102, 113, 117, 122, 125, 142
subRosa, 143–144, 153
success, 24, 59, 125, 130, 131, 159, 173
success, economic, 13, 17, 20, 49, 58, 66, 69, 91. *See also* entrepreneurship
success, scientific, 21, 100, 121
Summers, Lawrence, 181n3
Synthetic Biology Project, 28, 182n15

tantear (concept), 161
Taylor, Kyle, 34
teaching, 124–125, 127, 133, 173. *See also* education; pedagogy
technicians, laboratory, 14, 47, 130–131
technoscience, 57
terrorism, 74, 75–76, 78–79, 89, 181n4
terrorists, 12, 76, 78–79, 92, 93
Tetreault, Mary Kay, 127
theory vs. praxis, 161
tinkerers, 87t; and accountability, 59, 69, 109; and affect, 82; as antiracist, 68; as antisexist, 68; and Big Bio, 13, 62; and *bricoleurs,* 178n9; challenges faced by, 46–47; characteristics of, 11, 13–14, 58, 61, 148; colorblindness and, 21, 65, 68, 93, 94, 148; and concept of innocence, 63, 69; and concept of intelligence, 69; as co-opting social justice ideas, 88, 90, 93, 95; as counterculture figures, 62; and creativity, 11, 59; definition of, 6, 180n2; and democratic sciences, 10; demographics of, 42; and disenfranchisement, 21, 58, 61, 88, 90, 93, 94; DIY biologists as, 57, 90; and DIY biology communities, 30; and economic success, 69; education promoted by, 90; and embrace of failure, 121; and entrepreneurship, 58, 64, 93; and exclusion, 21, 58, 61; feminist, 148–149, 153–154, 156, 162; and funding for science, 62; and geeks, 65; and gender, 43, 59; genderblindness and, 21, 65, 68, 93, 94; and gendered racial capitalism, 58, 65, 93; and inclusion, 90, 148; and innovation, 62, 94; and knowledge production, 17, 57; and masculinity, 69; and nature (natural world), 57, 64; and nerd culture, 59; and passion for science, 63–64, 69; and play, 158–159; and politics of inclusion, 65; positionality of, 10, 24–25, 53, 61–63, 64–65, 93; as the public, 90; and race, 59; racialization of, 12, 93; and scientific ethics, 72, 90–91, 94; scientists as, 57, 58–59; and social class, 61–63; and structural power, 93; vs. terrorists, 12, 79; in United States political landscape,

11–12; white men as, 12, 64. *See also* amateurs; biohackers; biologists, DIY; *specific DIY biology groups*
Tinkerers: The Amateur, DIYers, and Inventors Who Make America Great, The (Foege), 11
tinkering: and *bricolage,* 178n9; colorblindness and, 16; definition of, 10; disability activism as, 163; and economic power, 13, 63–64, 68, 69, 91; and entrepreneurship, 13, 16, 49, 91, 151; ethics of, 150; and exclusion, 16; and feminism, 16, 142, 144–145, 162; feminist, 162, 172; and gender, 69; genderblindness and, 16; genetic research and, 16; Genspace and, 59; and innovation, 10, 57; and iteration, 166; and knowledge production, 53; and nature (natural world), 16; and neoliberalism, 16, 151–152; and objectivity, 57–58; as play, 16–17, 155; and positionality, 10, 165; and the public, 12, 13–14; and race, 69; vs. science as discipline, 30; and scientific ethics, 17, 93; social justice as, 147; and standpoint theories, 16; vs. traditional science, 148; in the United States, 10–11, 13, 69; and universities and colleges, 173; women and, 16. *See also* biohacking; biology, DIY; biology, synthetic
Tocchetti, Sara, 77
traditional science: DIY biologists and, 82
transfeminism, 123, 140–141, 152, 153, 165–166
transgender politics, 153
Transplant (Quimera Rosa), 165–166, 167
Traweek, Sharon, 40
Trump, Donald, 4, 12, 120
Trumpism, 11–12
truth: absolute, and science, 7; and affect, 113, 114t; and Big Bio, 178n5; challenges to idea of, 120; and decoloniality, 19; emotions and, 107; and knowledge production, 15, 55, 99, 105; and passion for science, 114t; and positionality, 24, 25; and science, 56, 100, 112, 177n5, 182n18, 183n6; scientists and, 57
Tyfield, David, 14

United States: capitalism in the, 54, 171, 178n6; democracy in, 171; DIY biology labs in the, 79; politics of the, 11–12; racism in the, 170; and reproductive justice, 152–153; terrorism and the, 76; tinkering in the, 10–11, 13, 69; underdog narratives in the, 61, 69; and the "war on terror," 89; workers in the, and globalization, 12 universities and colleges: and capitalism, 8–9, 14, 32, 34; and collectivity, 173; and concept of intelligence, 56; diversity in, 181n10; vs. DIY biology, 67, 81; feminist science in, 102, 131; feminist science studies and, 109; funding for science in, 183n3; gatekeeping by, 32, 83–84; and meritocracy, 159–160; people of color in, 8; and politics of inclusion, 8, 181n10; scientific ethics in, 70–71, 73; success in, 159; systemic inequalities in, 159–160, 172–173; and tinkering, 173. *See also* education; laboratories, traditional; pedagogy; sciences, academic; teaching

Ureña, Carolyn, 156–157

vaccines, 124

violence, 6–7, 8, 19, 132, 133, 156, 157, 171

virtue signaling, 42

Vora, Kalindi, 66

"war on poverty," 147, 166

"war on terror," 79, 89

Washington, Harriet, 123

Weasel, Lisa, 142, 164

Weiner, Charles, 85

Weinstein, Matthew, 143, 172

Whatley, Mariamne, 103, 128

whiteness, 9, 40, 63, 69, 76, 77, 79, 141

white people, 12, 42, 111, 158, 170

"White Privilege: Unpacking the Invisible Knapsack" (McIntosh), 105

white supremacy, 56, 100, 108, 158, 178n6

Wilding, Faith, 143

Willey, Angie, 59, 104, 105, 168

Willis, Hyla, 143

women, Asian, 42

women, Black, 9

women, Latina/o/x, 154

women, white: and coloniality, 131; in DIY biology communities, 42; and family structure, 184n6; and heteropatriarchy, 7; and intersectionality, 149; and labor, 65; liberal, 177n4; in science, 100, 123

women: and capitalism, 6–7; and COVID-19, 170; in DIY biology communities, 42, 43–44, 51; and gendered racial capitalism, 93; and gender essentialism, 177n5; inclusion of, 139–140; and intersectionality, 100, 149, 157–158; as irrational, 7; as partners to male scientists, 40; and passion for science, 41, 110; and privilege, 131; and the public, 87; as reproductive labor, 178n6; resistance to science by, 115; in science, 97, 99–100, 108, 116, 122, 123; and science illiteracy, 115; tinkering and, 16; in underpaid professions, 125; and United States conservative politics, 12. *See also* gender; genderblindness; sexism

women of color, 19, 132, 157, 185n3

women's health, 139, 141

women's health movement: and control of knowledge, 148; and definition of the human, 149; knowledge production by the, 144, 145, 148; and neoliberalism, 139, 166; and politics of inclusion, 150; and race, 149; and science, 56; and self-determination, 8; and self-help clinics, 144; and standpoint theories, 145–146

women's liberation movement, 116

women's studies: and accessibility of science, 129; and biological determinism, 128; and feminist science, 113; and feminist science studies, 102–103; feminist scientists and, 103–104, 106, 117; resistance to science in, 113; and science, 99, 101, 109, 125, 127; and science literacy, 113; scientists and, 106. *See also* feminist science studies; feminist studies

Woolf, Virginia, 131

Wynter, Sylvia: and definition of the human, 18; and idea of *Homo economicus,* 15, 65; and knowledge, 143; and liberation movements, 11; and science, 51, 108, 121–122, 150

Young Lords, 139

About the Author

SIG/SARA GIORDANO is an activist-scholar who works as an associate professor in the Department of Interdisciplinary Studies at Kennesaw State University in Georgia and specializes in feminist science studies. Dr. Giordano received their PhD in neuroscience from Emory University and previously worked as an ethics consultant for the Centers for Disease Control and Prevention (CDC). Their areas of interest are the politics and ethics of science, with a special focus on critical science literacy and the democratization of science. As a feminist scientist with previous bench science experience, they have the perspective of both a practicing scientist and a critical science studies scholar.